FIVE MEDIEVAL ASTROLOGERS

AN ASTROLOGICAL MISCELLANY
TRANSLATED AND EDITED
BY

JAMES HERSCHEL HOLDEN, M.A.
FELLOW OF THE AMERICAN FEDERATION OF ASTROLOGERS

First Printing 2008

ISBN-10: 0-86690-578-2
ISBN-13: 978-0-86690-578-7

Published by:
American Federation of Astrologers, Inc.
6535 S. Rural Road
Tempe, AZ 85283

Printed in the United States of America.

TABLE OF CONTENTS

PREFACE TO FIVE MEDIEVAL ASTROLOGERS

This volume contains five medieval astrological tracts. The first is the *Book of Flowers* by the famous 9th century Arabian astrologer Abû Ma<shar, better known as Albumasar. It is an interesting compendium of Mundane Astrology drawn from his larger works.

The second work is the well-known *Ptolemy's Centiloquy*, a book of 100 astrological Aphorisms relating to Natal, Horary, Mundane, and Electional Astrology. It was not written by Claudius Ptolemy, but its true author is unknown.

The third work is perhaps the next best known—the *Centiloquy of Hermes Trismegistus*. It too contains 100 Aphorisms relating to Natal, Horary, and Electional Astrology. It is one of many astrological tracts attributed to Hermes, but its true author is unknown.

The fourth work is the lesser-known *Centiloquy* attributed to the Arabian astrologer Bethen (or Bethem), but it is now thought to have actually been written by the 12th century Jewish astrologer Abraham Ibn Ezra (1092/3-1167). If so, it was originally written in Hebrew, then translated into Old French, and from that into Latin.

The fifth work is perhaps the least known today. Its full title is *The Propositions of Almansor to the King of the Saracens*, but it is more commonly known simply as *The Propositions of Almansor.* It is half again longer than anyone of the three *Centiloquies*, since it contains 150 Propositions or Premises. It was perhaps written by a late 9th or early 10th century Arabian astrologer al-Manṣûr of the Munajjim Family.

All five of these tracts were translated into Latin by the so-called 12th Century Translators and were among the earliest books printed in the late 15th and early 16th centuries. Thus they were all read and cited by the early modern European astrologers, but their popularity waned with the decline of astrology after the 18th century, and except for Ptolemy's *Centiloquy*, they became hard to find.

However, with the recent upsurge of interest in medieval and classical astrology, the time has come to make all these works available in a convenient collection. I translated Albumasar's *Book of Flowers* more than twenty years ago and circulated it privately, but this is its first publication. The *Centiloquies* by Ptolemy, Hermes and Bethen were translated and published in the 17th century by Henry Coley (1633-1707) in his book, *Clavis Astrologiae Elimata* (London: Ben Tooke and Thomas Sawbridge, 1676) pp. 315-345. Some of them have been reprinted from time to time.

I originally intended to use Coley's translations of the *Centiloquies* by Hermes and Bethen by modernizing their language, since many 21st century readers find 17th century English to be difficult. However, I later noticed that Coley had abbreviated some of the Aphorisms, particularly those of Bethen, so I have instead made new translations of those two *Centiloquies*.

In these translations, I have retained the older astrological terms 'figure', 'impedite', 'evil', and 'under the Sunbeams', and I have also expanded the translations in some places to make them clearer, since the Latin text is sometimes terse, particularly in Bethen's *Centiloquy*. In addition, I have added comments to some of the Aphorisms.

I have used Wright's translation of al-Bîrûnî and Kunitzsch's *Glossary* for explanations of the astrological terms derived from Arabic, most of which are now obsolete.

Since we now have a critical edition of the Greek text of *Ptol-*

emy's Centiloquy by Emilie Boer, I have made a new translation of that work from the Greek. The reader who has the 17th century translation of that *Centiloquy* made from the Latin by Coley, or the early 19th century translation made by J. M. Ashmand, will find some noticeable differences in the translations. In the case of *Almansor's Propositions*, I have also made a new translation of the Latin version.

A question that naturally arises in a modern reader's mind is, Why should I read an astrological tract written a thousand years ago? Perhaps the best answer is that it enables you in effect to attend a lecture by a skilled astrologer of the past, in which he gives you a large number of tips on how to read a chart, and also, from his own personal experience, he mentions some particular configurations to look for. I think that anyone who has the patience to read these tracts will find some helpful information. And I hope that these translations will prove useful to 21st century astrologers.

James Herschel Holden
Phoenix, Arizona
22 January 2008

ALBUMASAR

THE BOOK OF FLOWERS

TRANSLATED FROM THE 12TH CENTURY
LATIN VERSION OF JOHN OF SEVILLE
BY
JAMES HERSCHEL HOLDEN, M.A.
FELLOW OF THE AMERICAN FEDERATION OF ASTROLOGERS

FIRST EDITION 1986, DALLAS, TEXAS
CIRCULATED PRIVATELY

SECOND REVISED EDITION 1995, PHOENIX, ARIZONA
CIRCULATED PRIVATELY

2/Five Medieval Astrologers

PREFACE TO THE FIRST EDITION

The most famous of the astrologers writing in Arabic was Ja'far ibn Muḥammad Abû Ma'shar al-Balkhî (c.787-886), known in the West as Albumasar. He was a native of Balkh in what is now northern Afghanistan, but he spent most of his life in Baghdad. Al-Nadîm says that he died at al-Wâsiṭ on Wednesday 28 Ramadan A.H. 272, which is equivalent to 9 March 886, when he was more than 100 years old.

According to the Arabian historians, Albumasar was first a religious scholar who specialized in the adîth, the traditional sayings relating to the message of the Prophet Muḥammad. We are also told that he did not take up the study of astrology until he was 47, although this seems too late.

Coming as he did from Balkh, he was steeped in what remained of the pre-Islamic Persian culture, and, when he took up astrology, he also became interested in the astronomical theories of the Hindus. Thus, he had knowledge of astronomy and astrology from Greek, Persian, and Hindu sources.

Greek astrology had passed to the Arabs in the 8th century mainly through the translation of Greek books into Arabic, although some of the books seem to have been first translated into Syriac or Persian and later retranslated from those languages into Arabic. The Arabs bought Greek books in Constantinople and other cities of the Eastern Roman Empire and brought them back to Baghdad, where a special translation bureau had been established to produce Arabic versions of them.

In Baghdad, Albumasar either knew personally or had access to the works of such astrologers as Mâshâ˃âllah, Omar Tiberiades, and Sahl. And, with the excellent library and book shop facilities of that city, a great deal of traditional astrological knowledge was at his disposal. His own books show that he took advantage of these resources.

He was one of the court astrologers to the Khalifs of Baghdad and also evidently maintained a private practice. He mentions from time to time the theories of others and also the results of his own experience. His books covered the whole field of astrology. The most notable are his *Great Introduction to Astrology*, *The Great Conjunctions*, *Revolutions of the Years of Nati*vities, *Revolutions of the Years of the World*, and a brief compendium of mundane astrology *The Book of Flowers*.

The Book of Flowers contains rules for interpreting what is now called "the Aries Ingress," and thus covers the same subject as the *Revolutions of the Years of the World*. However, they are not identical.[1] In the first half of the 12th century, several of Albumasar's books were translated from Arabic into Latin by the industrious John of Seville. John's version of *The Book of Flowers* was printed at Augsburg in 1488 by Erhard Ratdolt. It sold out quickly and was reprinted the following year and a third time in 1495. Melchior Sessa also published an edition of *The Book of Flowers* at Venice (in 1488?). And there were two more printings in Venice in 1509 and 1515.[2] I have translated Ratdolt's edition of 1488, but I have added a few readings from a 13th cent. MS in the Bibliothèque Nationale (Parisinus lat. 16204) and some conjectures of my own. Since a few MSS of the Arabic text are still extant, it would be desirable to have a translation direct from the Arabic. Hopefully someone will do this eventually. However, Ratdolt's edition has some intrinsic interest, since it was used by many renaissance astrologers.

---

[1]David Pingree's statement in his article on Abû Maʿshar in the *Dictionary of Scientific Biography* (New York: Charles Scribner's Sons, 1970-1980) that the two books are the same shows that he had not examined them. Evidently his list of Albumasar's books must be used with caution.

[2]See Francis J. Carmody, *Arabic Astronomical and Astrological Sciences in Latin Translation* (Berkeley and Los Angeles: Univ. of California Press, 1956).

The order of the chapters in Ratdolt's edition differs considerably from that of the Parisinus, and there are numerous differences in the text. Since I have not examined any other MSS of the Latin version, I am not able to say which sequence is correct. I have generally followed the printed text, and I have only resorted to the Parisinus where I detected a problem. Thus, I have not taken full advantage of that MS, which has numerous differences from Ratdolt's edition.

In translating from the Latin into English, I have produced as literal a translation as seemed possible without creating an unreadable version. John of Seville evidently proceeded in the same way, so the reader of my translation will see some of the characteristics of the Arabic original. It would have been easy to modernize the language, but Albumasar lived in a world different from ours. The customs, concerns, and language of the 9th century Khalifate obviously do not coincide with those of the 20th century United States. If we were to conjure up a picture of Albumasar, he would be dressed in baggy pants and a turban, not in a modern business suit. Therefore, I have changed his words into English, without, I hope, having too much altered his style.

Modern astrologers will find the book interesting and will, I think, be astonished at the detail. Admittedly, many of the prognostications refer to conditions in open country in the Middle East, and an urban American astrologer is not likely to be concerned about locusts in the field or worms dropping into fruit trees, but here in Dallas I have occasionally seen the air dark and obscured (with dust), and it has not been too many years since Fig Newtons vanished for a season from the supermarkets because the figs (imported from India) were found to be full of worms.

As with all astrology, the predictions offer us a picture of the culture that produced them. Some of the conditions of a thousand years ago are altered, but most remain, even though somewhat transformed. The king may now be a dictator (perhaps calling himself "president" or "chairman"), but he is still an autocrat. The highwaymen are now highjackers. The nobles may still be nobles, but more often simply members of the ruling elite. The rich, the "untitled" (whom I take to be the "middle class"), the poor, the rustics (farmers and agricultural workers), and the merchants are still with us. Their relations are usually, but not always, less boisterous than in the old days. Religion is still disturbed by innovators and dissidents within, by the change in relative strength among the existing sects, and by the occasional rising of new sects. Perhaps the French are right when they say, "The more it changes, the more it remains the same."

I believe this is the first work of Albumasar to be translated into English. Hopefully, there will be others. His books are a valuable resource for astrologers and for those interested in the history of astrology.

James H. Holden
Dallas, Texas
April 28, 1986

PREFACE TO THE SECOND EDITION

My principal reason for issuing this second edition is to improve the appearance of the book. The text of the first edition was printed on a 9-pin dot matrix printer, and the Preface was printed later on a 24-pin printer. The result was readable but unattractive. This edition is printed on an ink-jet printer in Times Roman type.

I am still not aware of the existence of any other translation of *The Book of Flowers* into a modern language. My own translation, made from a microfilm, was begun on 5 August 1960 and completed on 10 November 1968. During the eight years from start to finish I only translated a few pages in any one year and was not quite halfway through in 1965. In addition to being professionally employed, I had several other spare-time projects under way, and I usually worked on the translation at rare intervals as a means of relaxation from something else. The translation remained in pencil for 17 years until I transferred it to a desktop computer in 1985 using the Allwrite word processor. In 1994 I converted the Allwrite text file to a WordPerfect document file. Later that year I corrected some typographical errors and revised the translation and the notes. I also changed the numbering of the notes from continuous numbering to page numbering.

On rereading the Preface to the first edition, I see that I failed to explain the reasons for my limited use of Parisinus lat. 16204. Here is what I had in mind but did not express adequately. First, my translation was completed before I acquired the microfilm of the MS. Second, the text given in the MS has so many variants from that of the printed edition that I was obliged to choose between accepting the latter as the basis of my translation or undertaking the more formidable task of first reediting the Latin text and then translating the new edition. But in order to reedit the text properly, it would have been desirable to acquire microfilms of two or three other MSS and collate them. This would have been an agreeable undertaking if I had nothing else to do, but that was not the case. Accordingly, I decided to stick with the text of Ratdolt's edition and confine my editing to those places where the text seemed obviously at fault. After all, my main intent was to make an English translation of the Latin version most commonly used by the 16th and 17th century astrologers, which was Ratdolt's edition.

Now it seems appropriate to supply some additional information about Albumasar and his literary output. And first the question of his birth date.

Al-Nadîm tells us in the *Fihrist*[3] that Albumasar died at al-Wâsiṭ, Iraq, on Wednesday, 29 Ramadan 272 A.H. (equivalent to 9 March 886) after he had passed the age of one hundred. Hence, he was born prior to 2 March 789 (A.H. years are shorter than A.D. years).

Pingree thinks Albumasar was born at 9:55 PM on 10 August 787, since he found a horoscope for that date in Book 3, Chapt. 1, of Albumasar's *The Revolutions of the Years of Nativities*.[4] However, while the date of the chart may be acceptable, Albumasar's

[3]*The Fihrist of al-Nadîm* ed. and trans. by Bayard Dodge (New York & London: Cambridge Univ. Press, 1970).

[4]See his edition of the Byzantine Greek translation *Albumasaris de revolutionibus nativitatum* (Leipzig: B.G. Teubner, 1968), pp. 126-128. The text begins "We should therefore make an example to demonstrate the arc of direction and the measuring and the assigning of both the rulership of the lots and the dodecatemories. Someone was born in the 4th clime in a city whose latitude was 37°; and the horoscope [ASC] was Taurus 2°54′, and the Moon in the same [sign] 12°43′. . . and we shall both make the nativity and arrange the aspects in it, etc." Note that the latitude of the city is given as 37°. The latitude of Balkh is 36N46, which seems to agree well. However, the Arabs may not have known the latitude of the city with any precision. Al-Battânî, for example, gives the latitude of Balkh as 41°, and he gives the latitude of Herat and Esfahan as 37°. So, it seems likely that the chart in question was not set for Balkh, but perhaps for Herat, Afghanistan, or Esfahan, Iran.

pupil and assistant Abû Saîd Shâdhân states in his *Sayings of AbûMa<shar* that his master "had no record of his [own] nativity but had made a universal interrogation. Virgo was the horoscope [ASC] of the interrogation, while the Moon was in Scorpio diametrically opposed to the sun and configured with Mars. And such a figure signified epilepsy."[5] This statement would appear to preclude the possibility that the chart noted by Pingree is that of Albumasar.

Another argument against the 787 date is that al-Nadîm says it was reported that Albumasar did not learn about the stars until he was more than 47 years old, and Shâdhân mentions some astrological judgments made in 829, when Albumasar was a court astrologer to the Caliph al-Ma>mûn (786-833). If al-Nadîm's statement is correct, then Albumasar must have been born not later than 783.

It should also be mentioned that Albumasar was a prolific writer. Al-Nadîm lists more than 30 titles of his books, and Pingree[6] extends this to 41, of which 15 or more are lost. Albumasar was known to the Byzantines, and some of his books were translated into Greek. Several were translated into Latin in the 12th century. A correspondent in Cairo, Mr. Farouk Elhiddiny, tells me that a few titles have been printed in Arabic in the Middle East and are currently available in that language.

As was mentioned in the Preface to the first edition, the order of the chapters in the Parisinus differs considerably from their order in Ratdolt's edition (which is followed in the present translation). I have not seen any of the Arabic MSS that contain *The Book of Flowers*, so I do not know in what order the chapters appear in their original language. Here, however, is a description of the text as it appears in one 13th century MS:

---

[5]See Lynn Thorndike, "Albumasar in Sadan," *Isis* 45, pt. 1, no. 139 (May 1954):22-32, which contains a translation of selections from a Latin version of Shadhan's *Sayings of Abu Ma<shar*. My citation is from the second paragraph of p. 28.

[6]See his article on Abû Ma<shar in the *Dictionary of Scientific Biography*, which gives a detailed list.

The *Book of Flowers* in Parisinus 16204.

[The corresponding chapter numbers of the present translation are added in square brackets.]

p. 333, col. 2 *inc*. Oportet te primus scire dominum anni . . . 'You ought first to know the Lord of the Year . . .'

Chapt. 1. Finding the Lord of the Year. [1]
Inc. "You ought first to ..."

Chapt. 2. What the Lord of the Year will Signify, and First about Saturn . . . [2]
Inc. "And when you know the Lord of the Year ..."

Chapt. 3. Impediment and Fortune and of the Other Planets from the Lord of the Year . . . [3]
Inc. "A chapter about the signification of things ..."

Chapt. 4. The parts of the Latitude of the Planets in the Twelve Signs and their Rising and Setting [both when] Direct and Retrograde. [11]
Inc. "A treatise on the parts of latitude ..."

Chapt. 5. What will Happen due to the Action of the Fixed Stars in the Revolution of the Year.[9]
Inc. "The knowledge of the fixed stars . . . "

Chapt. 6. What the Head and Tail [of the Dragon] will Signify and the Stars that are called Comets. [12]
Inc. "What the Head and Tail [of the Dragon] will signify ..."

Chapt. 7. The Knowledge of the Revolution of the Quarters of the Year and when it is Necessary to Revolve them and when it is not.
Inc. "And know that the year, when it is . . ."

Chapt. 8. On the Specific Action of Certain Planets in the Revolution.
Inc. "A chapter on the knowledge of the heaviness or lightness . . . " [4]

(other subheads):

"The Conjunction of Saturn and Mars, when they are
Conjoined in the Signs . . ."[8, part]
"A Treatise on Earthquakes of the same Planet, and
on Rainstorms . . ." [8, part]
"The Knowledge, moreover, of Rains . . ." [4]
"The Knowledge, moreover, of Fires in the Air ..."
"If, moreover, Mars is in Gemini . . ." [8, part]
"A Chapter on the Knowledge of Battles and War . . ." [6]

p. 353, col. 2 [*des.*] . . . *que collecta st in eo ex floribus & secretis astrorum si deus voluerit* '. . . which are collected in it from the flowers & the secrets of the stars, if God wills'.

In addition to his large book on mundane astrology *De magnis coniunctionibus* 'The Great Conjunctions', Albumasar wrote a second small treatise: *The Revolutions of the Years of the World* or *The Book of Experiments*. This is another short book on mundane astrology, but it is different from the *Book of Flowers*. In the Parisinus it begins: *Dixit Albumasar. Scito horam introitus solis in primum minutum arietis et* . . . 'Albumasar said: Know the hour of the entrance of the Sun into the first minute of Aries, and . . . '. It is half again as long as the *Book of Flowers* and contains different material, including an Aries Ingress for the date 18 Ramadhan 35 A.H. (19 Mar 656 A.D.) at about 7:20 PM LMT (assuming the coordinates of the place to be 45° E and 32°15′ N).[7]

Some writers have confused these two books, but they are entirely separate works.

James H. Holden
April 29, 1994
Phoenix, Arizona

[7]Calculation with modern parameters will put the Sun in 2°01′ or 2°02′ Aries at this date and time. However, Albumasar evidently used tables that were based upon a fixed zodiac. According to his tables, the longitude of the Sun would have been 0°00′ Aries. Theoretically, it would appear that the tropical zodiac should be used for Aries Ingresses, since any fixed zodiac is an arbitrary one, while at zero Aries in the tropical zodiac, the Sun moves from the southern hemisphere to the northern hemisphere—something definite and significant!

PREFACE TO THE PRINTED EDITION

An opportunity has now come to issue a printed edition of this work. It is essentially the second edition of 1994 with a few minor changes. The principal change is that the footnotes are now numbered continuously. I want to thank Kris Brandt Riske, the Executive Director of the American Federation of Astrologers, for preparing the text for publication.

ALBUMASAR
THE BOOK OF FLOWERS

This is the book that Albumasar collected from the flowers of those things which the superiors signify in inferior matters, and what is in the Revolution of Years, Months, and Days; and he was bringing this book along, just as on journeys because he put in it the flowers of things and the rest that he had picked out and that was pleasing to him.[8]

[1. FINDING THE LORD OF THE YEAR.][9]

First, you ought to know the lord of the year; and the knowledge of this thing is known from the hour of the entrance of the Sun into the first minute of the sign Aries. Therefore, you shall know the ASC in that same hour as most certainly as you will be able; verify the cusps of the twelve houses of heaven because error falls in this if it is neglected. And when you have done this, look at the lord of the ASC with the rest of the planets—the one who then has more strength from the testimony of the circles of the angle.

And whichever planet you have found in the ASC, the 10th, 7th, or 4th angle, afterwards the 11th [and] 9th, finally the 5th,

[8]This is the rubric given in the Parisinus. It is either a translator's or compiler's note and not part of the original text. The term 'flowers' is used to mean 'choice selections'; in other words, this is a book of excerpts from other books—an *anthology*. What is meant by the phrase "... and he was bringing this book along..." is not clear, unless we are to understand that Albumasar packed many useful things into it as a traveler would when packing his bag for a trip. The first printed edition of the text has simply: *Incipit tractatus Albumasaris florum astrologie* 'Albumasar's treatise *The Flowers of Astrology* begins'.

[9]I have added some chapter and subsection titles and provided numbers for the chapters.

[that is the one sought]. And you shall not prefer the MC to the ASC, nor the 5th house to the 9th, but let it be done following the aforesaid scheme.

And if you have found a planet in the ASC, you shall not seek another of those which were in other houses.[10]

Likewise, if a planet was not in the ASC, and there was [one] in the MC, you shall not look at the rest of the houses.[11] Likewise, if a planet was not in the MC but in the 7th, you shall not look at the rest of the houses. Likewise, if not in the 7th, but in the 4th, you shall not concern yourself with the rest of the houses.[12] And the one whom you have found in these houses, that very one will be the dispositor of the year if it has any dignity, namely of domicile or of exaltation, triplicity, term, or face.

If moreover there was in the ASC a planet with no dignity in it, and there was in the MC a planet having [dignity of] term and face, [then] because it has doubled its dignity, it is the one that is sought; and you shall not seek another.

And if there was [one] after them in the 7th, to whom is joined [the dignity] of domicile, exaltation, term [and] triplicity or face, it is the [one] sought.

[2. WHAT THE LORD OF THE YEAR SIGNIFIES IN THE TRIPLICITIES.]

[And first about Saturn.]

Saturn, therefore, found [as] the lord of the year, if he was in Aries or its triplicity, there will appear in those states over which one of those signs presides that is in the division of the East[13] some rich men, powerful men, and commanding men, who will do

---

[10]The printed text has 'other houses of the planets', but 'other houses' will suffice.

[11]Again the text has 'houses of the planets'.

[12]Again the text has 'houses of the planets'.

[13]The Aries or fire triplicity was considered to rule the eastern direction.

things through subtle talents and showy arrangements and display of marvelous things and prodigies without any haste, nay, with silence and beautiful appearance in them.

But if Saturn is well disposed in the house, he will signify what I have said, namely patience, pleasantness, and inquiry into things in general, of the public, and of the rustics, and their obedience towards the king along with their submission to him and to the extension of his voice, namely of his rule, and his loftiness.

And if he was badly disposed, he signifies the death of the wealthy and envy and hatred of them, accusations and slightings, with the hatred of the rustics towards them.

And if Jupiter was with him, their action will be in that which I have said, with silence and religion and display of faith, with justice and observance of their preceptors.

If, moreover, Jupiter aspected him from a good [position], there will come upon the rich who dwell in regions that are in the division of Aries or of its triplicity gifts and services from those rich [men] who are in the division of the region of that sign in which Jupiter was, and there will come to their assistance from the same parts every good thing in which they will rejoice.

But if there was an aspect of Jupiter to Saturn by opposition, there will be, instead of gifts and services, fear and contention from those states of the region that are in the division of Saturn toward the rich of those states that are in the division of Jupiter, or, on the contrary, toward those rich [men] who are in the division of Saturn, seeking war and surrender from them. And this will be according to the number of degrees of the aspect that was between Saturn and Jupiter up to the hour in which was the war and the contrariety.

And if there was an aspect of Saturn to Jupiter by square aspect, in place of wars there will be disagreement in things between those who were in the partitions of Saturn and Jupiter.

If, moreover, in place of Jupiter, it was Mars, join to him the nature which I have expounded to you concerning the nature of Saturn [but] with haste in the actions of the rich who are in those regions with much celerity and contemplation and a paucity of piety, pity, and gentleness.

But if his aspect was by trine or sextile aspect, there will come upon the rich who are in the division of the sign Aries and of its triplicity, from the rich who are of the division of the sign in which Mars is, gifts and services, joy and rejoicing. Likewise, in general, it will be good for their rustics and their states. And they will receive sustentation from those states that are in the division of the sign Mars is in. And things will be offered from those states of the natures of things that are appropriate to those states. And I have already made this plain to you in the larger book that I edited *On the Natures of Climes and States*[14] and that which is applicable to them from the villages and natures[15] of those states.

But if Mars aspects Saturn by opposition, in place of donations and services there will be evil, such as quarrels and enmities and contrairities among the rich of the regions of the sign where Mars was. And these rich [men] will go to those others to subdue them, so that they may be subdued and driven out. Likewise, the things that were carried out of those regions will be torn away and destroyed. And its citizens will be bad for the citizens of those regions.

Because if there was an aspect of Mars by square, there will be diversity in place of fear and a cutting off of property and of sought-after things from those parts. And war will not be appearing along with this.

---

[14]This may refer to the *Book of the Natures of Places and the Generation of Winds*, No. 35 in the list given by David Pingree in his article on Albumasar (Abû Ma<shar) in the *Dictionary of Scientific Biography* (New York: Charles Scribner's Sons, 1970-1980).

[15]The printed text has *ex villis et naturis* 'from the villages and natures', which doesn't seem to make sense. Perhaps we should read *ex illis naturis* 'from the particular natures' or something of the sort.

And if it was the Sun instead of Mars, with this there will be that which I said about the nature of Saturn with regard to the rich—a seeking after a kingdom and showiness and a seeking after celestial things.

But if he aspected Saturn by trine or sextile, there will come upon the rich who are in the division of Saturn gifts and services from the rich who are in the division of the sign in which the Sun was. And good and salable items are brought from those parts. And rumors will come upon the rustics from those parts making them joyful. And there will be that which is brought to the market in that same year of the substance of the Sun. And I have already explained this in the *Greater Book of Natures*.[16]

If, moreover, he aspected Saturn by opposition, instead of gifts and services there will be quarrels and war in the searching for those things that the Sun signified of showiness and kingdom in those states. And that which was produced from them of good to those states will be torn away. And that which came from joy-bringing rumors will be destroyed and converted into rumors terrible to hear.

And if the Sun aspects Saturn by square, there will not be open war, but among the rich there will be strife and diversity of counsel among them.

And if it was Venus instead of the Sun, with this there will be that which I said in regard to Saturn of the seeking of princes—seeking games and quietness and delight and songs and the appearance of all games, and the appearance of these things and their mainifestation, and to the citizens of those states among the rich and the rustics.

If, moreover, she aspected Saturn from a place of friendliness, there will come upon the rich who are in the division in which Saturn was gifts and services from the rich who are in the states that

[16]A lost book on the natures of the planets. Al-Nadîm refers to it as "Natures, a large book in five sections as AbûMa<shar divided it."

are in the division of Venus. For there will come from them everything good, and it will be that which is produced in that same time and because all of them are inclined to venerean things.

Further, if she aspected [Saturn] from a place of enmity, there will be, instead of gifts and services, wars and contrarieties. And these rich [men] will proceed to war against those rich [men]. And the good which was brought up from those states will be torn away. And the substance from those parts will be diminished. And wealth will be diminished. And there will come rumors, evil and foul. And all good will be converted into evil, and all joy into worry. And the root [cause] of all these things will be women and venerean things.

But if she aspected Saturn by square, war will not be appearing, but there will be a diversity in things between those rich [men] and the rustics and with the citizens of that same part, so that if there were a father in the states that are in the division of Saturn and a son in the states that are in the division of Venus, there will fall between them a diversity in all that which they apportion between themselves in the matter and in [their] business dealings and in other affairs. And similarly there will be among all men and [their] kindred that which I have said in regard to planets and their aspects in this book.

If, moreover, it was Mercury instead of Venus, there will be that which I said about the things of Saturn in the seeking of the rich—a seeking after grammar and fecundity of speech and philosophy and astronomy and medicine and dialectics. And the apparition of them in this time among the rich and the citizens of the regions which are in the division of Saturn.

And if Mercury aspects Saturn by trine or sextile, there will come upon the rich who are in the partition of Aries and its triplicity gifts and presents [and] everything good from the kings who are in the states that are in the parts of Mercury. And salable items and substance will be brought from there. And the substance of wealth from these things will be multiplied. And this will be praised in the states that are in the division of Saturn. And more of these things will be mercurial things.

But if he aspects Saturn by opposition, there will be, in place of gifts and services, contrariety and war. And these rich [men] will proceed to warring [against] those rich [men]. And whatever was brought forth of good will be torn away. And the substance from the same parts will be diminished. And property will be destroyed. And there will come rumors, evil and foul. And every good thing will quickly disappear. And every evil thing will appear from those states. This will be [the case] if Mercury was involved with the malefics, i.e. if he had any connection with them. And if he was configured with benefics in the hour of his connection, there will be less evil and less impediment; but his impediment will not prohibit the action of the opposition aspect. Know these things and understand it!

And if Mercury aspects Saturn by square aspect, war will not appear. And there will be diversity among the rich; and there will also be diversity between the rustics and the soldiers. And mercurial things will be the cause of this thing. And all this is plainly stated in the *Greater Book of Natures*.[17]

When in place of Mercury it was the Moon, then there will be that which I said in regard to the things of Saturn in the seekings of the rich, the apparition of legates[18] and examiners and the establishment of things for sale. And this will abound among the rich and the rustics at the same time.

But if the Moon aspected Saturn by trine or sextile, gifts and services will come upon the rich who are in the division of Saturn from those in the division of the Moon. And things for sale will be brought up from those states. And the property and goods from those things that are of the nature of the Moon will be multiplied, and they will be given over to those states in which Saturn was,[19]

[17]This same book is mentioned above in Chapter 2.

[18]Officials sent out by the king or ruler of a place.

[19]That is, those states signified by the triplicity in which Saturn was posited.

and there will be more of these lunar things.

And if the Moon aspected Saturn by opposition, in place of gifts and services, there will be war and contrariety, and these rich [men] will go to war against those rich [men]. And whatever was brought down from those parts will be torn away. And property will be destroyed. And they will find rumors evil and terrible, and every good thing will be frustrated and overturned and made evil.

And if there was an aspect of the Moon to Saturn by square aspect, war will not appear, and there will be diversity between the rich and the rustics, and contrariety between kinsmen because of lunar things.

The Earth Triplicity [which is] Southern.

When Saturn is lord of the year and in Taurus or its triplicity, there will be all that which I have said previously concerning Saturn among the rich and the rustics; and the good and evil that was in the part of the East will be in the part of the South in states that are in the division of the sign in which Saturn was in accordance with that which I have expounded to you.

The Air Triplicity [which is] Western.

When Saturn was in Gemini or its triplicity, there will be all that which I have said of good and evil in the western[20] part of the world.

The Water Triplicity [which is] Northern.

When Saturn was in Cancer or its triplicity, there will be all that which I have said of good or evil in the part of the North.

[Jupiter Lord of the Year.]

When Jupiter presides over the year and is its lord and descends

[20]The printed text has *septentrionis* 'northern' in error—probably a typographic error—the typesetter's eye having been attracted to that word at the end of the next caption.

into Aries or its triplicity, there will appear in those states over which he presides towards the East religion and quietude and the apparition of sects and the observance of preceptors with justice and good work. But if Jupiter is then in good state, it signifies a multitude of goods for the rich and the goodness of their hearts, and their joy, and the obedience of the rustics, and their esteem towards them in the same place. But if he is in bad condition, it signifies their contrariety.

And know that the action of Jupiter in the aspect of the planets to him is made just like the action in the case of Saturn, and the aspects of the planets to him equally.

But if Jupiter is in Taurus or its triplicity, the apparition of those events will be in the part of the South.

And if it is in Gemini or its triplicity, there will be an apparition of those things that I have said of the events in the part of the West.

And know that that which I have said to you about the action in the effect of Jupiter, that I might prove it, and that which I have said to you about the effect of Saturn is from the secrets of the knowledge of the stars in *The Revolution of Years*.[21]

[Mars Lord of the Year.]

If Mars is lord of the year and presides over it and is in Aries or its triplicity, there will appear in those states over which he presides towards the East evil and injury and rapine and a call[22] to war and flight in the exercise of the instruments of wars. And men will teach their own sons war and fighting and martial things.

But if Mars was in good condition in his own place, it will signify the king's victory over his own enemies and over those who contend with him, and the greatness of his own heart and his swift-

[21]This is another book by Abû Ma<shar, probably the *De revolutionibus annorum*, translated by John of Seville, and published at Augsburg by Erhard Ratdolt along with the *De coniunctionibus* in 1489.

[22]Reading *invitatio* 'invitation', 'call' rather than *imitatio* 'imitation'.

ness,[23] and the delight of the rustics with this, and their obedience towards him. And if Mars was in bad condition, it will signify a scarcity of victory over his own [subjects] and that they will conquer him, and a scarcity of stability over things useful and just, and his failure in everything that he works at.

And the action of the effects of Mars and the aspects of the planets to him is just like the work in the effect of Saturn and the aspects of the planets to him equally.

When Mars is in Taurus or its triplicity, there will be an apparition of those things that I have said to you of the events in the part of the South.

When Mars is in Gemini or its triplicity, there will be an apparition of those things that I have said of the events in the part of the West.

And if he is in Cancer or its triplicity, there will be an apparition of those things that I have said to you of the events in the part of the North.

[The Sun Lord of the Year.]

When the Sun is lord of the year, it signifies the glory and sublimity of the king, and the seeking after the loftiness of kings and all sublimities, and the display of utility and progress, love and fondness for divine things and a multitude of them, and among the rest of the men who belong to those very states that are in the division of the sign in which the Sun is, each one of them according to his own nature.

But if the Sun is in Aries or its triplicity, it will be in the part of the East. And if the Sun is in good condition in his own place, it signifies rectitude of the soldiers and their justice and their fitness

[23]The printed text has *magnitudinem cordis sue ac celeritatem et dilectionem rusticorum cum hoc*, which can be translated as above. But perhaps we should emend *celeritatem* to *caritatem* and translate the passage 'and the greatness of his own heart, and the esteem and the delight of the rustics with this'.

and victory over their enemies and over those who contend with them. But if he is in bad condition in the place, it signifies a scarcity of the apparition of riches in those very states . . . over their own enemies and the cause of their honor.[24] And if he is in Taurus or its triplicity, there will be an apparition of those things which I have said in the part of the South.

And [if] he is in Gemini or its triplicity, there will be those things which I have said in the part of the West.

If in Cancer or its triplicity, there will be an apparition of those things which I have said of the events in the part of the North.

[Venus Lord of the Year.]

When Venus is lord of the year and descends into Aries or its triplicity, there will appear in the states over which she rules a seeking after games and songs, and works of clothing,[25] and the instruction of men in that very time, and the desire of the rich and the rustics for this.

And if Venus is then in good condition, it signifies delicacy of the soul, and the joy and gladness of the rich and the rustics, and the obtaining of the region, and their victory with the enemies of those who contend against them.

Moreover, if she is in bad condition, it signifies the contrary of all these things.

If she is in Taurus or its triplicity, there will be an apparition of those things which I have said in the part of the South.

If she is in Gemini or its triplicity, those things will be in the part of the West.

If in Cancer or its triplicity, those things will be in the part of the North.

---

[24]Something seems to be missing where I have indicated a lacuna.

[25]Reading *indumentorum* 'clothing' instead of *instrumentorum* 'instruments'.

[Mercury Lord of the Year.]

When Mercury is lord of the year and is in Aries or its triplicity, there will appear among the rich of the states over which those signs rule in the part of the East a seeking after knowledge, namely of astronomy and of medicine and philosophy and of subtility in these [subjects]. And likewise among the rustics; and men will teach their own sons this.

But if Mercury is in good condition in the same place, it signifies beauty[26] and integrity among the rich, and the obtaining of these things, and victory over their enemies and the beauty[27] and integrity of their judges, bishops, abbots, and other such [persons], and traders in boys and concubines for the rich.

But if [he is] in bad condition, it signifies the contrary of all these things. And the action of the effects of Mercury through the aspects of the planets to him will be just as in [the case of] the action of Saturn equally.

And if he is in Taurus or its triplicity, there will be that which was said in the part of the South.

If in Gemini [or its triplicity], in the part of the West.

If in Cancer [or its triplicity], in the part of the North.

[The Moon Lord of the Year.]

When the Moon is lord of the year and is in Aries or its triplicity, there will appear in those states over which she rules towards the East a frequency of whispering, with a multitude of allegations[28] and rumors; and the rich and the rustics will delight in these things.

And if she is in good condition, it signifies the apparition of roads and paths, with the safety of travelers, and the attainment by

---

[26]This does not seem appropriate for Mercury.

[27]Again, *pulchritudinem* 'beauty' does not seem appropriate.

[28]Reading *allegationum* 'allegations' instead of *legatorum* 'legates'.

the wealthy of victory over their enemies.

But if she is in bad condition, it signifies the converse of all this. And the action of the Moon through the aspects of the planets to her[29] will be just as in the case of Saturn.

And if she was in Taurus or its triplicity, those things will be in the South.

If in Gemini [or its triplicity], in the West.

If in Cancer or its triplicity, those things will be in the part of the North.

[3. THE NATURES OF THE PLANETS.]

[Saturn.]

Saturn is significator of the rich, of old things, of religious persons, of farmers, and of old men. If he is impedited in the hour of the revolution, then whatever is his own indication of things is destroyed. If he is in human signs, it signifies that there will happen to them infirmities for a long time, and the decay of the rich, and there will be destruction of victuals, [and] the rich will become paupers. But paupers will die and magnates will be made sad. And there will be a drying up of the bodies of those in whose signs he is along with leanness and tertian or quartan fever, also flight and depression of the mind, want and fear of death, and the murdering of the wealthy.

If he is in a terrestrial sign, he signifies the destruction of those things that are in his division, such as a scarcity of seed, [and] damage to trees from worms falling into them like locusts.

And if he is in air signs, he signifies severe cold with many clouds, much heavy frost, and corruption of the complexion of the air, and thunder, and lightning flashes, and difficulty from many rainstorms.

[29]The printed text has *eum* 'him', but *eam* 'her' seems more appropriate.

And when he is in a water sign, he signifies hindrances in rivers and seas, [and] in shipwrecks; and animals of the waters will suffer. But if he is in good condition and not impedited, those things which I have said will be just the contrary.

If he is impedited in the hour of the revolution, judge destruction and hindrance.

And if he is fortunately placed in human signs, it is [as was] said above.

And if [he is] in terrestrial [signs], it signifies troubles and injuries from tremors and earthquakes, and the destruction of houses, cities, and country places.

And if he is in an air sign, there will be corruption of the air and its darkness, along with thunder and lightning, with fiery sparks and lightning running about in the air, and especially if he is impedited in air signs.

And if he is in water [signs], there will be hindrance in water, and shipwreck, overturning of ships at sea, so that travelers will be imperiled in the water, and animals of the waters will die.

But if in place of evil, there is fortune, change the judgment and say good in place of evil and security in place of fear. And look at the planet impediting him or at what sign the [Dragon's] Tail is in, because, if it is in a fire sign, the root of the hindrance will be from fires and combustions.

But if it is in a sign of wolves,[30] there will be hindrance from wolves.

Moreover, if he is in bad condition in a terrestrial sign, the root of the trouble will be from terrestrial things. [And] your narration will be similar in the rest of the signs.

Moreover, if a benefic aspects him in the hour of the revolution,

[30]Presumably, Aries or Scorpio, being signs of Mars.

judge strength and good.

Moreover, if a malefic aspects him, judge according to the one that has more testimonies, and you will not often be deceived by a weak effect, because nature makes its own effect unless it is rendered entirely weak in general. Know these things and work in accordance with them, and you will not err if God is willing.

[Jupiter.]

Jupiter is significator of nobles and judges, of bishops, and consuls, religious persons, good citizens, and sects. If he is impedited in the hour of the revolution, then all that will suffer hindrance which is peculiar to him.

If he is in human signs, it signifies the throwing down of the nobles and the wealthy, and the bad condition of their things, and the paucity of their donations, and the frustration of their class, with anxiety with respect to their own kin, also the destruction of the kingdoms of Babylonia and of the Arabs, and paucity of goods with much scarcity.[31] And this will be in those cities that are in the division of the sign Jupiter is in, along with the exercise of untruthfulness in the speech of men, with the display of evil and injuries and infirmities of body, with the debility of the acquisition of men.[32] If, moreover, he is in a terrestrial sign, it signifies the destruction of lands and paucity of the fruits of trees, of wheat and barley, and the fall of mildew upon the harvest.

If he is in an air sign, it signifies paucity of rain and corruption of the winds and of the air.

Moreover, if he is in a water sign, it signifies the destruction of those sailing on the sea, along with paucity of their acqui[si]tion, and the drying up of waters, and paucity of fish.

And if he is in a bestial sign, it signifies hindrance in animals, especially those of that particular sign that are used by men.

[31]Reading *caritatis* 'scarcity' rather than *curiositatis* 'curiosity'.

[32]That is, 'with their inability to acquire desired things'.

Moreover, if in place of evil [there is] fortune, change the judgment, and say good instead of evil, justice instead of injury, loftiness [instead of] dejection, [and] honor [instead of] disgrace. And look at the planet impediting him—in what sign it is, and join that one [in the judgment], and speak according to that which was expounded in the things of Saturn. And similarly, your narration will be in the configuration[33] of Jupiter with a fortune or an infortune.

[Mars.]

Mars, to be sure, [signifies] military advisers, and instigators of battles, and insurgents against the king, also sudden deaths with many infirmities, grave fevers, and the breaking off of roads, effusions of blood, accidents of combustion, and the drying up of river[s].

If he is in terrestrial signs, it signifies the destruction of trees through the combustion of heat, and strong injurious winds, and combustion of new crops in their own time.[34] If he is in an air sign, it signifies paucity of rain and severity of heat, and lightning, and injurious hot spells.

And if he is in a water sign, it signifies perilous shipwrecks of those at sea, [which happen] suddenly from winds blowing strongly.

And in a quadrupedal sign, it signifies hindrance in those quadrupeds that men use and which pertain to that sign. Afterward, look at the planet impediting him, in what sign it is, and blend your statement on this according to what was said in the things of Saturn. And your narration will be similar in the involvement of Mars with an infortune or a fortune, just as with Saturn.

[33]The text has *in iuuamento Iovis cum fortuna vel malo* 'in the aid of Jupiter with a fortune or an infortune', but the Parisinus has *in coniunctione iovis cum fortuna uel mala* 'in the conjunction of Jupiter with a fortune or an infortune'. Here, as elsewhere, 'fortune' and 'infortune' refer to 'benefic planet' and 'malefic planet' respectively.

[34]As from drought.

[The Sun.]

The Sun signifies magnates, the rich, and the honorable.

If he was impedited in the hour of the revolution, all that is his suffers hindrance, and infirmities will happen to the entire commonalty.

If he is in human signs, there will be hindrance in those things of it which pertain to men.

If [he is] in terrestrial signs, it will be in those things that are of the substance of the earth and of metals.

If he was in an air [sign], there will be hindrance in the air.

If [he was] in a water [sign], those [creatures] that are animals of the water will suffer hindrance.

And if [his condition is one of] fortune instead of a bad [one], turn the opinion around and say good in place of evil. And look at the planet impediting him as [I said] above.

[Venus.]

Venus is the significator of marriageable women. When she is impedited in the hour of the revolution, all that is hers suffers hindrance.

If [she is] in human signs, whatever of hers there is in men will suffer impediment.

If [she is] in terrestrial [signs, it will be in those things] that are of the substance of the earth.

If [she is] in air [signs, it will be in those things] that are of the air.

And if [she is] in water [signs, it will be in those things] that are of the substance of water.

And if [her condition] is [one of] fortune instead of evil, turn the opinion around and say good in place of evil. And look at the planet impediting her as [I said] above.

[Mercury.]

Mercury is the significator of writers, arithmeticians, businessmen, boys, and of the activities of school teachers.

If he was impedited in the hour of the revolution, all that which is his will suffer harm. And your narration will be on those things that pertain to him and on his portent in human, terrestrial, air, water, [or] bestial signs according also to the aspect of the fortunes and malefics to him and according to his involvement with them, just as is said in [the case of] Saturn.

[The Moon.]

The Moon is the significator of ambassadors and embassies, and of the generality of the common people, and their behavior and calm in every day.

When she was impedited in the hour of the revolution, she will destroy all that which belongs to her. And your narration will be on her and on all the things that are hers according to her presence in human, air, terrestrial, water, [or] bestial signs, according to the aspect of the malefics and the fortunes to her, and their involvement with her, [and] from which signs they aspect her in the same manner as was said of the others.

Know and understand this secret, and there is nothing that can be hidden from you that you wish to investigate, if God is willing.

[4.] HEAVINESS AND LIGHTNESS [IN THE PRICE] OF GRAIN.

You will know these things from one or the other of the superior planets, because all heaviness is from the action of Saturn and all lightness from the action of Jupiter. Therefore, whenever you have seen Saturn in the revolution of the years in the sign wherein was the Conjunction that Signified the Sect[35] and neither Jupiter nor Venus with the Dragon's Head aspected him, this will be a sign of heaviness without doubt. Therefore, mix together your narration on him and fear not. And if it happened that the [Dragon's] Tail was with him in one sign, judge heaviness of [the price of] grain and fear among the people unless Jupiter aspects this place in the hour in which the [Dragon's] Tail is separated from him.

Likewise, the years which signify famine are those which Saturn presided over in the conjunction or opposition in which the revolution was; and the more heavily if he was conjunct or opposite Mercury; this, moreover, is done in years.

For in the months when you have seen him in conjunction or opposition in the 9th house or the 3rd, this will be a sign of lightness.

Know, therefore, the ASC's of the full Moon or the conjunction; after this, look at its lord, who, if he was increased in light or course, then the price of grain will be increased in that same month. And if he was diminished in course, the price of grain will be diminished. And if he was going toward his own fall, similarly the price of grain will fall. But if he was in the angle under the earth,[36] or in the 7th, the price will stand where it is.

Moreover, in things besides wheat and barley, you will look at the dispositor of the year [to see] if he was fortunate. Also, look at what kind of sign he is in and what the substance of that same sign is. But if it was of the substance of fire, this will be in gold and silver and in everything that is accomplished through fire. But if he was in terrestrial [signs], this will be in terrestrial things. Moreover, if he was in an air sign, then this will be in airy things.

---

[35]This is a reference to the mean conjunction of Jupiter and Saturn that occurred in the year 570 A.D. in the sign Scorpio and was held to be the conjunction indicative of the Muslim sect, since it occurred at about the time of the birth of the Prophet Muḥammad. See Abû Mashar's *Revolutions of the Years of the World* and his *The Great Conjunctions*.

[36]That is, the 4th house.

Look at the place of the same sign from the ASC and speak upon that. Look if he was in water signs, [for then] this will be in animals of the water and in all that which goes forth from it. All these, moreover, are places in which is the whole weight of grain.37 Moreover, if it was in Aries or its triplicity, it will be in the Eastern part.

If [it was] in Taurus or its triplicity, in the Southern part.

If, moreover, [it was] in Gemini or its triplicity, it will be in the West.

But if [it was] in Cancer or its triplicity, it will be in the North.

Consider the market for grain when the Sun has entered the first minute of Aries or of the signs in which are the exaltations of the planets that are [also] movable signs.[38] For the descent of the Sun into the first minute of Aries is stronger and more subtle than all its [other] descensions into the beginnings of the movable signs. These are its descensions into the movable signs. Take note, moreover, of this secret of astrology, so that you don't neglect it when you are going to the market.

[5.] RAIN.

The knowledge, moreover, of this matter is from Mars. When he was in any of his own signs in the hour of the revolution, this will be the beginning of much rain. And when he was in a sign of Saturn, this will be a judgment of a scarcity of rain. But in other signs of the other planets, it signifies a moderate amount of rain.

[6.] BATTLES AND WAR.

The knowledge of battles and war is from the place of Mars in his conjunction with Saturn [or] from his opposition or square. For when he was [placed] thus, he signifies war. And when Jupiter was

---

[37]I don't know what he means by this last statement.

[38]He means the other three cardinal sign—Cancer, Libra, and Capricorn—each of which is also the exaltation of one of the superior planets.

with Mars, it signifies war, as if in defense of justice for those who say they are by this action imitating justice.[39] And similarly, the conjunctions of other planets whould be mixed in according to the quality[40] of their natures; and let the essence of the images in the signs be examined according to what we have said in the Book of Images.

[7.] PESTILENCE.

[The knowledge] of pestilence and death among the multitude is known from the ASC of anyone of the principal [indicators] previously mentioned, such as the sign of the conjunction of the current year or the sign of the conjunction which signified the Sect; these things, namely when they are impedited and [also] their lords, there will be a severe pestilence. And if some of them are safe,[41] there will be death and pestilence according to its quantity. Consider this and you will find that, and you will be able to judge death in half of the men or a third or a fourth part, and you will not err.

[8.] EARTHQUAKES.

Earthquakes and floods are known from Saturn when his rays are in the signs of the first conjunctions and their ASC's; i.e., in any of those signs previously mentioned which are principals according to the sign of the conjunction, i.e. the ASC of the year, and the sign of the conjunction of the Sect or its ASC. For when Saturn is in them or projects its rays into them by opposition or square aspect, and when this itself is impedited. Look at him. If he is in terrestrial signs, he will make darkness of the air and earthquakes and famine. In air signs, snow and cold, darkness of the air, and its corruption, [and] blasts of wind uprooting trees and palms. If [he is] in water signs, submersions and floods according to the quantity of his strength will these things be happening. If in a fire sign, he

[39]The printed text has . . . *hac causa iustitie imitare*, but should we read *initiare* 'initiating' for *imitare* 'imitating'?

[40]Reading *qualitatem* instead of *quantitatem* in the printed text.

[41]The printed text has *salva* 'safe', which presumably means 'not impedited'..

makes terrible prodigies in the air and horrible things that are in the air.[42] For these things are known from Mars when he is the ruler in those years already mentioned. Afterwards, it happens that in the same year in which he was, he is the ruler of the same sign which he ruled in the aforesaid principle [configurations]. Therefore, according to the quality of his strength in the same year will be the strength of those things that will happen of the aforesaid sparks[43] and prodigies.

And know that the year in which the ASC is in its own strength, in that same year the hour of the descension of the Sun into the first minute of Aries, i.e. when a fixed sign is ascending, you should not revolve the quarters of that same year, for its ASC will suffice for you for the year; the mark of this thing is the fixed sign. For it will not require a revolution of the quarters and months unless for a business properly besides the rest.[44] Know this, moreover: if it is a common sign, it is always necessary that you revolve the year at the hour of the descension of the Sun into the first minute of Libra.

But if the ASC of the year is a movable sign, it is necessary for you to calculate the entrance of the Sun into every quarter from the revolution [of that quarter]. And you shall work out these figures just as you worked in the beginning of the year—this is the time of the descension into the first minute of Cancer, and the first minute of Libra, and the first [minute] of Capricorn.

Know this, and work that out, so that you may find in everyone [of them] that you have worked out the ASC's of the quarters just as you have done in the revolution of the beginning of the year likewise. And you shall not pay any attention to that because I have studiously tested what the ancient wise men have said, and I have found a part [of it] to turn out and a part to be wrong, and this [is] on account of the diversity of the aspects, which is variable in every month. For they have also said that you do in the ASC's of

---

[42]Reading *aere* 'air' for *acre* 'sharp' in the printed text.

[43]Probably meteorites. *Cf.* Ptolemy, *Tetrabiblos*, iii. 13.

[44]The meaning of the end of this sentence is obscure.

its quarters and in the ASC's of its months like you do in the ASC of the year; because you work in the ASC of the year through the planet which is stronger by testimony and place.

Moreover, they said it is unnecessary for you to do similarly in the ASC's of the quarters, but you should look at the planet [that is] strong in its own place, not by an abundance of testimony, but by the goodness of the place.

Others said you should work with the lord of the conjunction or the full Moon which was before this. And if it does not aspect the ASC, there will be its own [kind of] trouble, particularly in states which are of its division. And do in the aspects of the planets to it just as you have done in the work of the planets to it in Aries and in its triplicity; and speak on the part in which there will be the hindrance because of those things in the substances of the signs in which are the planets aspecting the lords of the quarter just as I have demonstrated in the chapter on Saturn.

Moreover, if Mars is in Gemini or its triplicity and Mars is lord of the year and direct in an angle, it signifies war and the effusion of blood and contention because that triplicity is of the human likeness; if [it is] retrograde, it loses its strength and weight. But if it is not in the angles and is direct and aspects the ASC, it signifies infirmity from winds and blood. And [this will be] weightier in the region which is of the division of the sign in which it descends beyond the triplicities;[45] if [it is] retrograde, [it signifies] pestilences. And if it does not aspect the ASC and is direct, [it signifies] much burning and hindrances through fires and malign infirmities to men.

If Mars is in Cancer or its triplicity and is itself lord of the year and direct, wars and contention will happen in the land of the Arabs, and blood and pestilence and death in abundance, and fear will come over the king,[46] [and] mutation from region to region.

If, moreover, he is not in an angle and aspects the ASC and is

---

[45]The meaning of this is uncertain.

[46]The Latin has *timebitur super regem*.

direct, in this [case] it must be feared to speak on this; if [he is] retrograde, it signifies the weightiness of this thing. But if he does not aspect the ASC, this will be particularly in one place . . .[47] from the aspects of the planets to him is just as in the preceding chapter on Saturn and the aspects to him likewise.

The conjunction of Saturn and Mars in the signs, when they are conjunct in the ASC of the revolution of any year, they signify a universal misfortune among the rustics.

And if they are conjoined in the 2nd sign from the ASC, it is the destruction of their livelihood and houses, and the memory of riches is committed to oblivion; and their servants and soldiers will be arrogant and will retain little loyalty to them.

And if they are conjoined in the 3rd, they signify the destruction of houses of religion and a descent of horrible things on those worshipping God. But if they are conjoined in the 4th, destruction of buildings, mansions, and farms.

If [they are conjoined] in the 5th, much destruction among children and a changing of the hearts of friends.

In the 6th, it signifies hindrance pertaining to animals and all quadrupeds which men use, and to their male and female servants.

In the 7th, the fall of misfortune and quarrels between partners and spouses and contending among themselves and bewailing the end of enemies.[48] In the 8th, many wanderers and their death and illness and occupation, and in those lawsuits that are made against the estates of the dead; and death will fall upon those who wander in that year.

In the 9th, roads and estremities are destroyed, and more [especially] in those, routes by which one arrives at houses of oration, greater and famous.

---

[47]There seems to be a lacuna here.

[48]This last prediction seems rather curious.

In the 10th, the destruction of kings and of the rich and the death of a great king.

In the 11th, destruction falling among friends; and everyone having a friend above his own comrade will be changed.[49] In the 12th, hindrance descending upon all quadrupeds; and trouble will fall upon all sellers of beasts.

And everything that I have said to you about this will be in regions that are in the division of the sign in which they are conjoined. Also, if they are conjoined in the signs and in the term of any planet, then there will be troubles in the states of the division of the same planet whose term it is.

[9.] THE FIXED STARS: WHAT THEY EFFECT IN REVOLUTIONS AND NATIVITIES.

I shall explain to you the knowledge of the fixed stars and their effects. Know, therefore, that in the Head of Aries there are 2 stars: one in 13°25′, the other in 14°25′, and their latitude is north and their complexion of the complexion of Saturn and Mars.

The Pleiades are of the nature of Mars and the Moon, and they are in Taurus between 9°55′ and 13°, [and] their latitude is 6° N.

Aldebaran, whose nature is of the nature of Mars, is in 19°15′ of Taurus. [Most of the remaining stars are unidentified.]

In Gemini is a star [in] 18°55′ (in another book, 16°), whose latitude is south, and its complexion is of the complexion of the Sun.

In Cancer, there are 2 stars, of which one is 12°55′, of the complexion of Mars; the other, 2°55′ of the complexion of Saturn.

In Leo [there are] 2, of which one is 15°55′, the other 7°22′, and their latitude is north, of the complexion of Saturn.

In Virgo is a star of the complexion of Mars [in] 7°01′ (in an-

[49]That is, friendships will turn sour.

other book, 4°), and another of the complexion of Saturn [in] 25°.

In Libra is a star of the complexion of Saturn [in] 26°.

In Scorpio [there are] 3 stars, of which one is in 1°03′, another [in] 9°, another [in] 8°07′, and their complexion [is of the complexion] of Mars.

In Sagittarius [there are] 2 little ones, of which one [is in] 19°02′, another [in] 21°01′, another [in] 25°08′, and they are of the complexion of Saturn.

In Capricorn [there are] 2, of which one [is] evil [in] 27°02′, the other [in] 29°05′ of the complexion of Saturn.

[There is] one in Aquarius 9°04′ of the complexion of Saturn.

In Pisces [there is] one [in] 4°07′ (in another [book], 4°).

[10.] THE LORD OF THE YEAR: IN WHAT MANNER HE IS IMPEDITED BY THE STARS.

When, therefore, you see the lord of the year joined to any one of those stars which I mentioned, i.e., when you see him in any one of those degrees that we have defined, know that the impediment will fall upon kings and rich men and nobles. And among them there will be mourning and malign cogitations and infirmities from headaches and delirium.

But if any one of those fell in the ASC degree, [and] if anyone was born in that same year and his ASC was in those very degrees of that sign, that child will be bad and unfortunate [and possessed] of many infirmities, of which the majority will be in the head.

And if the lord of the 2nd from the ASC of the revolution of the year fell with those stars which I have said to you, know that the impediment will fall in houses of goods. And the goods of the rich and the nobles will be destroyed. And obscurity and loss of mind will fall upon them. And they will judge themselves to be paupers with mourning. And malign infirmities will happen to them in the neck and ears, such as catarrh and deafness.

And if any one of those fell in the degrees of the 2nd from the ASC of the revolution of the year, and a boy was born in that year, and the sign of the revolution is the ASC of the Native, and that same star was in the degrees of his ASC, he will be a pauper and of bad condition and with little earning-power and of small talent among his own kindred. And if he has any property, it will disappear. And he will have many infirmities of the throat and ears.

Moreover, if the lord of the 3rd from the ASC of the revolution of the year was joined to any one of those stars which I have said, know that the impediment and destruction will fall in the houses of speaking[50] and on a man who was properly a stranger. And dispute will fall among brothers. And everyone will be made sad with his own brother without cause. And there will be many severe and protracted infirmities of men [arising] from the shoulder-blades and the *furcula*[51] and the arms in that time.

And if any one of those stars fell in the degrees of the third [house] of the revolution, and a boy was born in the same year whose ASC is this third sign, and a star is in the degrees of his ASC, he will be born over his brothers, a bad man, and [over] his friends. And whoever has it will constitute a friend for himself of bad will, bad thought, [and] no good purpose. And if he travels abroad, he will have no good from this. And if any wealth happens [to come] to him, it will be lost. And he will have many infirmities in the shoulder-blades, the *furcula*, and the arms.

If the lord of the 4th from the ASC of the revolution of the year was with any of those same stars which I have said, know that the impediment and harm will fall on lands and harvests and farmers; and war will fall upon the cities and country estates, and dispute among fathers and sons with much envy, and they will be made

[50]That is, lecture halls and such like.

[51]The word usually means *wishbones* (as of birds), but is perhaps here used for *clavicles*.

sad, each in turn. And the last of those things that will happen in the same year at every bad thing.[52] And if anyone deserves an inheritance or country estates, he will not see anything that he will especially prize in them.

And similarly, if any one of those stars fell in the degree of the 4th house from the ASC of the revolution, and if a boy was born in the same year, he will be born entirely bad [in his relations] with his parents, holding them in hatred. And the last of the things will be in evil. And if he does any work, he will not be praised for it. And his infirmities [will be] many. And those infirmities that happen at that same time will be in the chest and lungs and those parts of the body.

And if the lord of the 5th from the ASC of the revolution of the year was with any of those stars, infirmity, destruction, and impediment will fall among loved ones, and those who gave will not be rewarded, and men will hold their sons in hatred and also their sons' requests. And sexual relations will be hampered. And there will happen to him infirmities in the belly and the stomach restraining them from intercourse. And similarly, if any one of those stars fell in the degrees of the 5th from the ASC of the revolution, and if a boy was born in that same year with those same degrees of that sign ascending, he will be hateful to his own parents, with little respect for those who will be joined together with him; he will not persevere in fondness for anyone, nor will he reward anyone for good. And his infirmities in the belly and stomach will be many.

But if the lord of the 6th from the ASC of the revolution of the year was with any one of those stars, infirmity and destruction and impediment will fall upon all quadrupeds, and upon [those persons] serving the king, and his slaves, and similarly upon the other slaves, and he whose business is the sale of slaves will be made poor, especially if the lord of the sign [is] in a masculine sign. Or this impediment will fall among maidservants and women if it is in

---

[52] A word or two has fallen out of the text.

a feminine sign. And fugitives will suffer in that year, and some of them will not return. And there will be many infirmities among men, and many of these in the intestines, and colics. And similarly, if any one of those stars fell in the degrees of the 6th [house] of the ASC of the revolution. And if a boy was born in that same year, and [the degrees of his ASC] are those same degrees of the ASC from that sign, he will not be lustful[53] among the slaves. And if he owns slaves, he will not see any good from them. And he will delight from childhood to flee from his parents. If he wants to use quadrupedal animals, he will not see any good from them, but they will die at his place. And there will be infirmities in him and in men [in general] at the same time in the lower part of the belly.

And if the lord of the 7th from the ASC of the revolution of the year is with any one of those stars, litigants and warriors will be multiplied over the king, and they will slay each other. And harm and quarrel will fall among men. And association will be cut off with much disintegration. And the king will be angered against his own familiars. And men will divorce their wives. And piety and pity will be diminished, and their hearts will be hardened. And few marriages will be made. And no profit will go to the seller. And investigations will be broadened, and capital-assets will be diminished. And men will have infirmities in the bladder and in the kidneys and the hips and the posterior parts. And similarly, if any one of those stars fell in the degrees of the 7th from the ASC of the revolution of the year, and if a boy was born in the same year under those degrees, he will not see any good from his own wife nor [from] anyone who is associated with him. And if he invested capital, he will not see what he wants in it. Contending with anyone, he will be defeated. And he will have infirmities in the bladder, kidneys, hips, and the posterior parts, as was said above.

But if the lord of the 8th from the ASC of the revolution of the year was with any one of those stars, infirmities, destruction, and impediment will fall upon wanderers. And no one will steal any-

[53]This word does not seem appropriate.

thing in that time that he will not return. And the king will miss something from his own treasury, and it will not be found. And impediment and many quarrels will fall among the heirs of the property of the dead, and much mortality among the lowly and dejected, and much impediment upon the properties of those contending with the major king of the regions. And they will need the aid of the greater king of the kings of Babylonia and the king of the Romans[54] and the Indians. And men will have many infirmities in the penis and genitals and testicles. And if any one of those stars falls in the degrees of the 8th from the ASC of the revolution, it will be done in the same manner as aforesaid for the lord of the 8th, if it falls with any one of them. And if a boy was born in that same year under those degrees, he will be a fugitive, and he will not enter under a roof, and he will be shaking hands with enemies, and he will try to be with demons. And there will not be seen in itself any good from his own labor and travel. And very often he will be in a place of the dead and of corpses and stenches. And many of his infirmities, but also [those] of men [in general] in that time, will be in the testicles and the private parts.

And if the lord of the 9th from the ASC of the revolution is with any one of those stars, impediment, destruction, and infirmity will fall upon those worshipping God, and upon the greater house of oration, and upon the ways of those traveling to far distant places. And the religions of men will be destroyed. And there will fall upon them complication and mental bewilderment and thoughts upon celestial things and doubt and denial[55] of the Lord God the Highest. And whoever travels abroad will not see good from his travel. And the morals of men will be destroyed, and their thoughts will be malicious. And rebels will go out and seek the death of the King of Babylonia. . . .[56] And this will be prolonged for them, that is, it will last until the year passes. And in that year there will be few battles and little slaughter and war. And likewise, if any one of

[54]The Greek emperor at Constantinople.

[55]Reading *negatio* 'denial' for *negotiatio* 'business' in the printed text.

[56]Something seems to be missing here.

those stars falls in the degrees of the 9th from the ASC of the revolution, and if anyone was born in that year under those degrees, he will be stupid in those things and in his own practice of the faith. And neither travel nor change will suit him. And his infirmities will be many, and at the same time those of men [in general], in the thighs and the muscles of the thigh.

And if the lord of the 10th from the ASC of the revolution is with any one of those stars, evil and destruction will fall upon kings, nobles, and the rich, and upon those who are in command. And envy and contention and weakness in their things,[57] and a small destruction[58] And their respect and their kingdom will pass away. And the commons will be extended over them. And their own rustics will blaspheme them, imposing upon them derisory names, to such an extent that if the chief of them should pass through the market place, he will be jeered at by the poorest of the merchants. And this will last until the year goes out. And they will have many infirmities in the knees.[59] And if a boy was born under that same degree of the ASC, he will be base and despicable among his own close kin and among those associated with him. And he will be of little respect and much dejected from a laborious life and have little gain from those things that persist to him. And he will have many infirmities in the knees.

And if the lord of the 11th was with any one of those [stars], anxiety will fall upon the hearts of enemies, and there will be more discord than usual among them. And one will suspect the other. And there will be little trust in anyone of them from those which he thought.[60] And the king's treasure house will be destroyed, and his wealth will perish. And he will have few attendants, and his troops will be disloyal. And if he is involved in war, they will abandon him. And men will have infirmities in the legs. And if a boy was

[57]That is, in their properties and power.

[58]Reading *deletio* 'destruction' rather than *delatio* 'accusation' in the printed text.

[59]Reading *genibus* 'knees' rather than *genitalibus* 'genitals' in the printed text.

[60]The printed text has the phrase *que putabat* 'which he thought', which seems to be corrupt.

born under those degrees, he will be of little faith and small good; and, hateful to all men, he will not avoid evil, nor will he intend good. And he will be of little praise and favor. And he will have infirmities in the legs.

And if the lord of the 12th was with any one of those stars, then bandits and highwaymen[61] will be multiplied. And slaves will be of no use to their masters, and the commoners will be diminished and despised [as well as] the nobles, magnates, wise men, and mean of reason. And if a boy was born in the degrees of the same ASC, he will be troublesome, crafty, [and] wicked with many enemies hating him. Everyone who sees him will hate him from his youth until old age. And he will have infirmities in the feet.

And if Jupiter or Venus aspect these places from friendship[62] and project rays to those same degrees and are in their own place or in their own houses, it will break down the malice of these [fixed] stars. And if they were weak in their own place, they will not perform anything of their own works that is effective. And the amount of evil of the planet will be according to the amount of its strength or weakness. And if it aspects from a place of enmity, it will not repel that evil, especially if it was weak in its own place, but it will strive to drive it out and won't be able. Moreover, if a malefic aspects from a place of dejection, the evil that I have mentioned will come with haste. And that evil follows another. And if a malefic aspects from a place of enmity and contrariety, there will be all that which I have mentioned from the beginning of the happening to its end. Moreover, if the Sun aspects the place, it will not hide the evil but will make it open. And everyone who tries to hide is unable [to do so]. . . .[63] but it will be for the whole generation and for cities. And if Saturn aspects, [there will be] one trouble after another.

And if Mercury aspects, there will be his effect just like the

[61]Literally, 'those who cut off roads'.

[62]Read 'from [a place of] friendship' or 'from [an aspect of] friendship'.

[63]Something seems to be missing here.

Sun's effect, but it will be less and somewhat quicker on account of the swiftness of his course.

And know that all that which was said will be in the cities that are [under the rulership] of the same sign. Therefore, consider this carefully, and you will not err if God is willing.

[11.] THE LATITUDE, THE RISING, AND THE SETTING OF THE PLANETS IN THE SIGNS.

You ought to consider the planets in the hour of the revolution and to make the latitudes in the parts of them into the north, the south, [and their positions with respect to the Sun] into the west and the east.[64]

[Saturn.]

If Saturn is in Aries and his latitude is north,[65] it signifies corruption of the air and thickness and darkness of it in its effects.[66] And if his latitude is south, it signifies severe cold with much frost.

And if he is oriental, it signifies death of the rich and of their relations[67] with sadness and frost.

If occidental, it signifies earthquake and famine and darkness of the air.

And if he is direct, there will be less evil on that account.

[But] if retrograde, more evil.

---

[64]This sentence is not at all clear as it stands. Either some words (such as I have placed in square brackets) have fallen out, or else it is simply an inept statement of the contents of the chapter.

[65]Since Saturn's north node was near the middle of Cancer in the 9th century, Saturn could not have north latitude in Aquarius through Gemini, nor south latitude in Leo through Sagittarius. Only in Cancer and Capricorn can it have either north or south latitude. Hence, the prognostications based upon the impossible latitudes must have been derived entirely from theory.

[66]So the Parisinus. The printed text has *...in tempore suo* 'in its own time'.

[67]Reading *cognationes* 'relations' for the *cogitationes* 'thoughts' of the printed text.

If Saturn is in Taurus and his latitude is north, it signifies goodness of the air and beauty of its physical condition. And there will be useful rains and without hindrance. And things bought and the market will be middling.

And if the latitude is south, it signifies alarms and mixing of weather and mortality and corruption of the air and its destruction, with little good.

If oriental, it signifies much rain, and men will be made infirm by this.

If occidental, it signifies fear among the generality[68] of the rustics and mixing of their things in turn. And if he is direct, it signifies infirmities happening to men in the upper parts of the body, and lasting.

If retrograde, it signifies the death of great men and nobles, with much fear and want among men.

If Saturn is in Gemini and his latitude is north, it signifies the blast of harmful winds and darkness of the air and earthquake and corruption.

If south, it signifies hot weather with little rain and much pesilential mortality.

If oriental, it signifies infirmity of the king and of the magnates.

If occidental, it signifies dryness of the air, with scarcity of water in springs, and [scarcity] of rains.

If direct, [the preceding indications are] stronger.

If retrograde, these things work weaker.

If Saturn is in Cancer and his latitude is north, it signifies scarcity of water in rivers and in rains.

If south, it signifies the engagement of men in their own activ-

[68]Literally, 'reach'.

ity and the scarcity of their increase.[69] If oriental, [it signifies] corruption of the air and its darkness and much cold in its own time.

If occidental, it signifies many rains and corruption and hindrance of things that are bought.

If retrograde, it signifies the strength of them, and the disgrace of the king, and his hindrance that will happen to him, than which death will be better.

And if direct, [it signifies] clarity of the air and goodness of its physical condition.

If Saturn is in Leo and his latitude [is] north, it signifies much rain and the seizing of the things of the rich and the rustics, and the increase of businessmen will be destroyed.

[If south, it signifies....][70] If oriental, it signifies infirmity falling upon the lowly and the untitled.[71] If occidental, it signifies pestilence falling upon the earth and mortality.[72] If retrograde, [it signifies] the seriousness of those things which I have said and the long continuation of their duration.

If direct, it signifies [that they will happen] more gently.

If Saturn [is] in Virgo and his latitude north, it signifies goodness of winds and the sweetness of their increases on harvests and seeds.

If it is south, it signifies scarcity of the waters of springs.

If oriental, [it signifies] abortion among pregnant women and other animals.

---

[69]I am not sure what Albumasar means by this.

[70]No signification for south latitude is given by either the printed text or the Parisinus.

[71]Lat. *ignobiles*, lit. 'ignobles, of low-birth, or base'. I have taken this to mean free men or non-aristocrats, so I have translated it as 'untitled'. But perhaps it is merely meant as a synonym of the preceding word *viles* 'lowly'. The phrase recurs several times below.

If occidental, it signifies deaths and terrible fevers.

If retrograde, it signifies the king's fear of his enemies, and he will fear their augmentation over himself.

If direct, it signifies the king's attainment of victory over his enemies and his strength.

If Saturn is in Libra and his latitude north, it signifies dryness of the air and its heat and scarcity of rains and wasting of the waters of springs.

If south, it signifies the goodness of winds and the sweetness of the physical condition of the air.

If oriental, it signifies the mingling of men and their indulgence in lewdness and crimes.

If occidental, [it signifies] the casting down of fornicators and their fall and [other] horrible things.

If retrograde, it signifies the infirmity of maid servants and men of low degree.

If direct, [it signifies] a middling amount of grain and barley, and the breaking of the wealth of kings.

If Saturn is in Scorpio and north latitude, it signifies much rain and an abundance of the waters of rivers and their corruption.

If south, [it signifies][73] famine for men and high prices for the market.

If oriental, many wars among the rich.

---

[72]The printed text has *mortem validam 'strong death'*. Guido Bonatti, *Tractatus de revolutionibus annorum mundi*, Chapter 64, has *mortalitatem* 'mortality'. I have adopted that reading. Cf. below, Cauda in Cancer, where the printed text has *multitudinem pestilentie & mortes subitaneas* 'a multitude of pestilence and sudden deaths', and Bonati has *multitudinem pestilentie & mortalitatis* 'a multitude of pestilence and mortality'.

[73]From this point forward, the printed text generally omits the phrase 'it signifies'.

If occidental, the destruction of abundances in the sea and the fall of evil among them.[74] If retrograde, pestilence in that land.

If direct, good circumstances in the clime of Babylon.

If Saturn [is] in Sagittarius and his latitude north, it signifies a abundance of water in springs and a great winter cold.

If oriental, the expulsion[75] of the rich from their own regions.

If occidental, war among the rich, the magnates, and the nobles.

If retrograde, high prices of the things of men and war among them.

If direct, prosperity from travel on the sea and the water.

If Saturn [is] in Capricorn and [his] latitude north, it signifies the goodness of the physical condition of the air and a middling amount of rains.

If south, darknesses of the air in wintertime with great cold.

If oriental, bad attitude among the rich and error and bad relations of them with the rustics.

If occidental, a multitude of locusts and the hindrance of harvests on account of too great heat.

If retrograde, much mingling among the rustics.

If direct, it signifies justice from the rich and the kings and their humbleness and rectitude.

If Saturn [is] in Aquarius and [his] latitude north, it signifies much rain [and] severe cold with frost.

---

[74]The meaning of this is not clear.

[75]The printed text has *proximitatem* 'proximity', which does not seem to make sense. Perhaps we should read *expositionem* 'expulsion'. *Cf.* William Ramesey, *An Introduction to the Iudgement of the Stars* (London: Robert White, 1653), p. 252, who has 'put out'.

If south, scarcity of the waters of springs and rivers.

If oriental, scarcity of the acquisition of inheritances, with scarcity of property among the rustics.

If occidental, much combustion and its strength.

If retrograde, seriousness of the same thing.

If direct, the death of animals that men use.

If Saturn [is] in Pisces [and his latitude] north, a multitude of blasts of north winds and severity of the winter.

If south, the breaking up of ships at sea and the fall of impediment on those that similarly work in the waters.

If oriental, mutual killing among the rich.

If occidental, dejection of the nobles and exaltation of the untitled.

If retrograde, death of the religious and of those worshipping God.

If direct, ravaging of houses of religion and of divine worship.

[Jupiter.]

Jupiter, when he is in Aries and his latitude north,[76] signifies fitness of the physical condition of the air and scarcity of rains.

If south, corruption of winds and their severity.

If oriental, it signifies joy and happiness among rich men and their inclination to play.

---

[76]Since Jupiter's north node was near the end of Gemini in the 9th century, Jupiter could not have north latitude in Capricorn through Gemini, nor south latitude in Cancer through Sagittarius (except for the first 10 degrees of Cancer and Capricorn and the last 12 degrees of Gemini and Sagittarius, where it can have either north or south latitude). Hence, the prognostications based upon the non-occurring latitudes must have been derived from theory.

If occidental, sublimity of the wise and the nobles.

If retrograde, hindrance of navigation at sea.

If direct, a multitude of fish and animals of the waters, and their health.

If Jupiter [is] in Taurus and his latitude north, it signifies stillness of the air and its heat.

If south, abundance of rains and an abundance of water in springs.

If oriental, hatred of the rustics towards their own king.

If direct, the desire of men for treasures and the collecting of wealth.

If Jupiter [is] in Gemini, it signifies fineness of the winds and a middling amount of rains if [its] latitude [is] north.

If south, much thunder and lightning.

[If] oriental, much discourse in sects and in the faith.

[If] occidental, many wars and disputes and minglings.

[If] retrograde, the high price of purchases.[77] [If] direct, the opposite of all these things.

If Jupiter [is] in Cancer and latitude north, [it signifies] fineness of the air and abundance of waters and much thunder and lightning.

[If] south, beauty of the air and its physical condition.

[If] oriental, much war and strife among regions mutually.

[If] occidental, much contention among sects.

[If] retrograde, severity of all those things that were said.

---

[77]The Latin has *gravitatem acquisitionis*.

[If] direct, safety in the mountains and their harvests.[78]

If Jupiter [is] in Leo and latitude north, it signifies severity of thunder and lightning.

If south, dryness of the air and scarcity of waters.

If oriental, mutual disagreement among the rich.

If occidental, it signifies that the rustics will subdue their own king.

If retrograde, a taking away of the hands of the soldiers in obedience to their king.[79] [If] direct, increase of birds and cows and camels.

If Jupiter [is] in Virgo and latitude north, [it signifies] scarcity of thunder and lightning and much rain.

[If] south, beauty of the air and its physical condition.

[If] oriental, much warfare and battles among the rich.

[If] occidental, contrariety of the soldiers and those bearing arms towards the rich.

[If] retrograde, severity of those things that I have said about this.

[If] direct, the religion of kings and their fitness.[80]

<If Jupiter is in Libra and its latitude is north, it signifies many westerly winds.

If south, a scarcity of rains.

If oriental, it signifies impediment that happens to men from wolves.

---

[78]This sounds wrong, since we don't normally think of harvests in the mountains. Perhaps the word 'mountains' is corrupt.

[79]This apparently refers to punishment by mutilation. *Cf.* Jupiter oriental in Aquarius below.

[80]This apparently means that kings will be religious and fit to rule.

If occidental, it signifies the wrath of the king towards his rustics.

If retrograde, it signifies the malice of the king's law.

If direct, it signifies the chastity of the king in his law.>[81]

If Jupiter [is] in Scorpio and latitude north, [it signifies] fineness of winds.

[If] south, much thunder.

[If] oriental, the fall of hindrance upon the nobles.

[If] occidental, much pestilence among mortals.

[If] retrograde, severity of all these things.

[If] direct, obscurity of the air and the fall of hindrance from this.

If Jupiter [is] in Sagittarius and its latitude north, [it signifies] temperateness of the air and its fineness.

If south, scarcity of rains.

If oriental, mutual disagreement among the rustics.[82] <If occidental, it signifies the appearance of lascivity.

If retrograde, it signifies much falsehood [on the part] of the young.[83] If direct, it signifies the mutual fitness of the rich .>[84]

If Jupiter [is] in Capricorn and latitude north, it signifies much thunder and lightning and many rains.

[If south, . . .][85] If oriental, it signifies the sublimity of the

---

[81]The significations of Jupiter in Libra are missing from the printed text. I have supplied the omissions from the Parisinus.

[82]The Parisinus has '. . . it signifies diversity among the rustics and the common people in sects'.

[83]Lat. *minorum mendacii.*

[84]The translation in angle brackets is supplied from the Parisinus.

[85]The signification of south latitude is missing in both the printed text and in the Parisinus.

sects.[86] If occidental, honor to the wise men and the doctors of law.[87] If retrograde, hindrances to the magistrate[88] and the scribes.

If direct, fitness of them and of other men.

If Jupiter [is] in Aquarius [and latitude] north, [it signifies] darkness of the air and scarcity of rains.

If south, much thunder and lightning.

If oriental, the cutting off of hands and feet in that time.

If occidental, much mingling among the rustics.

If retrograde, the severity of all those things that I have said.

If direct, fitness of the nobles and their sublimity.

If Jupiter [is] in Pisces [and latitude] north, it signifies an abundance of waters in rivers and springs.

If south, much rain and a scarcity of thunder.

If oriental, [it signifies many] wars [and battles].[89] If occidental, pestilence <in that same time>.[90] If retrograde, infirmities happening to men in the lower parts of the body.

If direct, mutual mingling of men.

[Mars.]

Mars in Aries and latitude north[91] signifies heat in the air and scarcity of rain.

---

[86]The printed text has "... sublimity of the stars, i.e. their goodness." I have preferred the reading of the Parisinus.

[87]The Parisinus has "... wise men and religious men."

[88]Lat. *consulum*, 'consul', but here probably a judge or magistrate.

[89]The Parisinus has ... *significat multitudinem guerre et bellorum*, where the printed text has only ... *bella* 'battles'.

[90]The Parisinus adds the words in angle brackets.

[91]Since Mars's north node was around 11 Taurus in the 9th century, Mars could not have north latitude in Capricorn through Pisces, nor south latitude in Cancer through Virgo. In Aries through Gemini and Libra through Sagittarius it can have

If south, severity of thunder and lightning.

If oriental, severity of wars among the rich,

If occidental, much terror among men.

If retrograde, infirmities in the eyes and head.

If direct, much illumination in faith and temptation.

If Mars [is] in Taurus and latitude north, it signifies many rains and the fitness of plants.

If south, many northerly winds.

If oriental, much peace and security in the East.

If occidental, it signifies many infirmities and death.[92] If retrograde, infirmities falling among boys.

If direct, the hatred of women.

If Mars [is] in Gemini and latitude north, it signifies many rains.

If south, scarcity of waters in rivers and springs.

If oriental, many blisters among men.

If occidental, the concealment of consuls and scribes and their flight.

If retrograde, disagreement among [religious] sects.

If direct, much poverty among men.

If Mars [is] in Cancer and his latitude [is] north, it signifies scarcity of waters in springs with much cold in its own time.

If south, [a heavy] blast of winds and the destruction of trees,

---

either north or south latitude. Hence, some at least of these prognostications must have been derived from theory.

[92]Reading *mortis* 'death' instead of *motis* in the printed text.

If oriental, pestilence will fall among animals that are moved about in the fields.

If occidental, [seditious] murmuring of the commons along with treachery.

If retrograde, the appearance of fornication and lechery.

If direct, fineness of the air and beauty of its physical condition.

If Mars is in Leo and his latitude north, it signifies scarcity of waters in springs and rivers.

If he is south, it signifies an abundance of water in springs and rivers.

If oriental, impediment in animals that men use.

If occidental, scarcity of fish and the death of aquatic animals.

If retrograde, sadness of the nobles and the rich.

If direct, joy among the judges and among those who know laws.

If Mars is in Virgo and his latitude [is] north, it signifies that men [will be] hindered in harvests and sowings.

If south, the salvation[93] of men in harvests and sowings.[94] If oriental, much death among the old.

If occidental, dryness and famine in Spain and its parts.

[If] retrograde, much war among men.

[If] direct, honorable behavior of the rich.

If Mars is in Libra and his latitude [is] north, it signifies darkness of the air and severity of winds.

---

[93]By *salvationem* 'salvation' he presumably means the reverse of 'hindrance', but the sense is obscure. *Cf.* Mars in Sagittarius, occidental, below.

[94]Reading *sementibus* 'sowings' in place of the *seminibus* 'seeds' of the printed text.

If south, much thunder and lightning.

If oriental, war among the rich.

If occidental, you will inscribe[95] great anger among kings and the seizing of the rich.

[If] retrograde, severity of those things that I have said to you.

[If] direct, many wars among the rich.

If Mars is in Scorpio and his latitude [is] north, it signifies many clouds, rains, [and] much thunder and lightning.

If south, an abundance of water in springs.

If oriental, many infirmities among men in the lower part of the belly.

If occidental, scarcity of piety among men mutually.[96] [If] retrograde, much good in that time, and dread among businessmen, and much gain among them.

[If] direct, a multitude of reception[97] of riches and the devastation of houses and the property of the rich.[98]

If Mars is in Sagittarius and his latitude is north, it signifies beauty of the physical condition of the air and its goodness.

If south, profit among businessmen and goodness of their gain.

If oriental, peace and quiet.

If occidental, the good condition[99] of trees and an abundance of fruit.

---

[95]That is, 'you (the astrologer) will write [in your interpretation]...'

[96]I don't know what he means by 'mutually'.

[97]The word 'reception' seems inconsistent with the rest of the sentence; perhaps 'seizing' was what Albumasar had in mind.

[98]Or, '...and devastation of the houses and property of the rich.'

[99]The printed text has *salvationem* 'salvation', but the sense requires *salubritatem* 'health, good-condition'. Evidently John of Seville considered the noun *salvatio* to encompass all the meanings of the verb *salveo*, one of which is *salvus sum* 'to

[If] retrograde, much coughing and pain in the chest.

[If] direct, much disease on land and among the animals of the seashore.

If Mars is in Capricorn and his latitude [is] north, it signifies much snow and cold.

If south, heat of the air and its obscurity.

If oriental, the murder of the Emperor of the Romans[100] and the salvation of Al-Kahizhir.[101] If occidental, many pestilences among men.

[If] retrograde, a high price for grain.

[If] direct, an abundance of grape-gathering and of [edible] oil.

If Mars is in Aquarius and his latitude [is] north, it signifies much snow and cold and an abundance of locusts in its own time.

If south, obscurity [of the air] and severity of heat.

If oriental, peace, and joy among the soldiers.

If occidental, much trembling among men.

[If] retrograde, heat, and impediment [falling] on trees.

[If] direct, the destruction of trees and the falling of worms among them.

If Mars is in Pisces and his latitude [is] north, it signifies finess of the air and of its physical condition.

If south, an abundance of locusts and a scarcity of hindrance to them.

---

be in good health'. But this is not proper Latin, nor, for that matter, proper French, Spanish, or Italian. Curious!

[100]The Byzantine emperor.

[101]I cannot identify this word, unless it is a corruption of *al-Khalîfah* 'the Caliph (of Baghdad)'.

If oriental, mutual murder among the rich.

If occidental, many infirmities among servants and maid-servants and men of the lowest class.

[If] retrograde, good health[102] among men and much gain from business.

[If] direct, the good condition[103] of sheep and cows.

And know that I have already carefully tested the latitudes of the superior planets and the degrees[104] and their being in the circle of signs.[105] Moreover, I saw the [things indicated by the] inferior planets sometimes to come to pass [and] sometimes to err. And therefore I have not put them in my book. And I have already carefully tested all these things that I have put in it. Know this. And you may not neglect anything of the things that are written in it. And judge on this what you wish, and you will not err, if God is willing.

[12. THE MOON'S NODES, AND COMETS.]

Know that Caput,[106] when it is in Aries, signifies the elevation of the great and the nobles and dejection of the lowly and the untitled.

If Cauda[107] is there, it signifies a bad arrangement among kings and [actions of theirs causing] injuries to the rustics.

And if any one of those stars called *Comets* appears there, it sig-

---

[102]The printed text has *salvationem* 'salvation'. See the note to Mars in Sagittarius, occidental, above.

[103]Again, the printed text has *salvationem*.

[104]The printed text has *partes*, which I have translated as 'degrees'.

[105]This seems to mean that Albumasar has verified to his own satisfaction that the latitudes of the superior planets do not depart from the zodiac, which was assumed to be a belt extending to 8 degrees of latitude on either side of the ecliptic.

[106]That is, *Caput Draconis* 'The Head of the Dragon'—the North or Ascending Node of the lunar orbit.

[107]That is, *Cauda Draconis* 'The Tail of the Dragon'—the South or Descending Node of the lunar orbit.

nifies the destruction of the rich, of the East, and of Babylon, and a bad existence and much sadness among the rustics.

And if its appearance is in the east, it will be swifter in its effect.

If in the part of the west, it will be slower.

If Caput is in Taurus, it signifies the murder of the rich in the region of the North, and much disagreement between the rich of the West and the rustics.

And if Cauda [is] there, it signifies a scarcity of piety in the hearts of men, and much travel in the region of the North, and scarcity of gain for them.

And if any one of those stars called comets appears in its direction, it signifies evil to men, and little good for them, and injuries [arising] from rebels towards them.

But if it appears in the part of the east, its effect will be swifter.

If in the part of the west, slower.

If Caput is in Gemini, it signifies infirmities among men [arising] from winds, and earthquake, and the fall of wars between the rustics and the rich.

If Cauda is there, it signifies an uprising of the rustics against the king and the nobles. And they delude them. His soldiers rise against them, and dukes [rise] against kings.

But if any one of those stars called comets appears in its direction, it signifies the appearance of lewdness and fornication, and the dejection of those serving and praying to most holy God.

But if it appears in the east, its effect will be swifter.

And if in the west, slower.

If Caput is in Cancer, it signifies the goodness of the king toward the rustics, and his largess to them, and that he will collect wealth; and he will scatter about more than he collected in his own places.[108] And if Cauda [is] there, it signifies much pestilence and many sudden deaths and false stupors, and the appearance of luxury, and the scarcity of justice, and the destruction of wealth,[109] and the destruction of warehouses, and much travel among the rich, and their removal from place to place, and the hindrance of scribes and to their masters.[110] And if the comet star appears there, it signifies a multitude of locusts, and the loss of harvests from them, and the fall of worms into wheat and trees, and scarcity of fruits, and many worms in them.

But if it appears in the east, its effect will be swifter.

If in the west, slower.

If Caput is in Leo, it signifies an abundance of lightning and fires in the air, and a very great [amount of] slaying, and great evil and injuries.

And if Cauda is there, it signifies much obscurity of the land, and earthquake, and the increase of waters, and the diminution of trees, and destruction of harvests.

And if the comet star is there, it signifies troubling from wolves, and the hindrance of men by them, and the fall of worms into wheat, and the destruction of warehouses.

And if it appears in the east, its effect will be swifter.

And if in the west, slower.

If Caput is in Virgo, it signifies destruction of the harvest and of other fruits, and scarcity of what is harvested, and the descent of vermin onto trees, and destruction by combustion of those things

[108]An unusual occurrence in the Middle Ages, but commonplace in the 20th century.

[109]The printed text has *fractionem census*, literally 'breaking of the census'; the Latin word *census* can mean 'wealth' or 'property', and I have translated it in that sense, but I may be wrong. Perhaps the text is faulty.

[110]The printed text has *et dominis suis* 'and to [or, from] their masters'. This doesn't seem to hang together too well with what precedes. Something seems to have dropped out.

that remain in them in the time of its custody.[111] And if Cauda is there, it signifies hatred falling among the rich; and in turn they will go against themselves in war, and captivities will be multiplied, and plunderings of the church and of the house of greater prayer, and the fall of diversity among men in speech and faith.

But if the comet star is there, it signifies the taking away of certain ones of the domestics of the rich, and their changes into other regions with a certain amount of captivity and evil, and their goods will never be returned.

And if it appeared in the east, its effect will be swifter.

And if in the west, slower.

If Caput is in Libra, it signifies injustices of the king and of the rich in general towards their own rustics, and deceptions and searches for that which is not among them and not in their hands. And on account of this, plundering will fall upon some and poverty upon others. And [it signifies] suffering among certain ones and delusion and punishment.

But if Cauda is there, it signifies the falling of death among cattle; and that year will be dry, with severe cold in its own season and heat in its season destroying crops and plants and seeds and trees, and [it also signifies] a scarcity of gathering fruits.

And if the comet star is there, it signifies a spreading out of bandits, and cutting-offs in parts,[112] and the destruction of wealth,[113] and the appearance of poverties[114] in the hearts of men in various fashions, i.e. sometimes with cold and sometimes with heat, and this will be prolonged among men.

And if it is in the east, its effect will be swifter.

---

[111]I don't know what he means by 'custody'.

[112]I'm not sure what Albumasar means by this—perhaps the isolation of certain regions.

[113]Again, the phrase *fractionem census* occurs. *Cf.* Cauda in Cancer above.

[114]The word 'poverties' does not seem to make sense here.

If in the west, slower.

If Caput is in Scorpio, it signifies an abundance of abscesses[115] among men and of blisters among their magnates, and many wars and battles, and many fornications among women, with eviction and the falling of them into the hands of kings, along with their mourning and sadness.

And if Cauda is there, it signifies the falling of delirium among men, and infirmities happening in [their] chests, and a multitude of catarrhs in the throat, [but] at the same time the joy and delight of the rich, and the gathering up of wealth, and plundering of their houses.[116] But if a comet appears there, it signifies many wars and battles, and the revolt[117] of kings, and the change of soldiers over them, and inquiries by them with impossibility.[118]

And if it is in the east, its effect will be swifter.

If in the west, slower.

If Caput is in Sagittarius, it signifies the affliction of the rustics by the King of Babylon, and the severity of its injury [to them], and hindrance in sheep, and cattle, and horses particularly, and in every animal that warriors use. And ruin will fall upon instruments of war. And the air will be corrupted and obscured. And heat will be made greater in its own time.

But if Cauda is there, it signifies the dejection of the great and the nobles, and the exaltation of the lowly and the untitled. And doctors of the law, wise men, scribes, and counsellors will be saddened. And hidden impediments will come upon them.

---

[115]The printed text has *abstematum*, for which read *apostematum*. But Bonatti has an entirely different prognostication: "It signifies good [conditions] and joy among men of the middle class, but bad [conditions] and sorrow among the magnates...."

[116]The "plundering of their houses" seems inconsistent with "the joy and delight of the rich."

[117]Or perhaps 'the overthrow'.

[118]I don't know what he means by this last phrase.

And if a comet appears there, it signifies a bad situation for the scribes and the doctors of the law, and the destruction[119] of their property, and the covering up of them with much trouble that will descend upon them.

But if it is in the eastern part, its effect will be swifter.

If in the west, slower.

If Caput is in Capricorn, it signifies joy among the rich and the nobles, and the loftiness of the great, with exaltation of these same, and the falling of tails (?),[120] and the hindrance of them, with dejection of the untitled and their ruin.

If Cauda is there, it signifies earthquake and mortal wounds.

If a comet [is there], it signifies the spread of fornication among men in that time.

If Caput is in Aquarius, it signifies the death of doctors of the law in that time, and of judges and followers of the faith.

And if Cauda is there, it signifies the king's searching out of the lords of inheritances in that which is not destined for him, and the taking away of their unregistered things, and injuries and afflictions, along with [de]frauding of them.

But if a comet is there, it signifies an abundance of war and killing and much affliction in that time.

And if it appears in the east, its effect will be swifter.

If in the west, slower.

If Caput is in Pisces, it signifies loftiness of the nobles and the advancement of each man in his own order more than he will advance according to the quantity of his own nature, and the gathering up of assets in houses of assets.

---

[119]The printed text has *antelationem* 'anticipation', but Bonatti has *anullationem* 'destruction', which makes more sense.

[120]Either the text is corrupt or something is missing here.

And if Cauda is there, it signifies the increasing of [the power of] the soldiers over the king (and similarly, over military commanders), and the moving of very many of the rich from their own dwellings, with an abundance of speech in the [established] sect and the appearance of new false ones.

If a comet is there, there will be a very great war upon the relations of the king. And they will kill themselves in turn, and they will be killed by others who will be elevated above them; and [it also signifies] the tearing away of their hands from obedience.

But if its appearance is in the east, its effect will be swifter.

If in the west, slower.

Moreover, when you want to know the region in which that which the comet signifies will happen, look at the zenith of the tail, in what region it is. And in that same region will be that which it signifies of trouble. The hour, indeed, of the remaining effects (which I have said to you in this book), in which will be all that which has been said, will be when it comes to the sign that is the substance of those things that I have said to you or to the planet whose nature is [involved] in the effect, just as the nature of the event, which [planet] appears to you in the same year.[121] This is [one] of the secrets of astronomy[122] because it is worthwhile to be hidden.

[Colophon of the Latin translation.]

The work of the *Flowers* of Albumasar ends happily. [Produced] by the exceptional diligence of that ingenious man Erhard Ratdolt of Augsburg and by the marvelous art of printing, in which, lately at Venice, now at Augsburg, he excels most notably. The 14th day before the Kalends of December 1488. [Hence, 18 November 1488.]

[121]This long, awkard sentence seems to mean that the time of the event is indicated by a progression or transit of a significator to the sign or planet indicative of the event.

[122]He means "astrology."

PTOLEMY'S CENTILOQUY

TRANSLATED FROM THE GREEK
BY JAMES HERSCHEL HOLDEN, M.A., FAFA.

TRANSLATOR'S PREFACE

This tract, which is called *Karpós* 'Fruit' in Greek, was certainly not written by Claudius Ptolemy, but who did write it and when is unknown. According to Emilie Boer, the first confirmed mention of it is by the Syrian bishop Severus Sebokt in the seventh century, but whether he had it in Syriac or Greek is not known. It seems to have been known in an Arabic version in the ninth century,[123] and there is a Latin version that was translated by John of Seville from an Arabic version with an extensive commentary by Haly ('Alî ibn Riḍwân, d. 1068); it was first printed by Erhard Ratdolt at Venice in 1484.and reprinted by Bonatus Locatellus at Venice in 1493. I incline to think that the original version of the *Centiloquy* was in Greek, and that it was subsequently translated into Arabic by some unknown astrologer who added comments of his own (since the Arabic version is more wordy than the Greek version).

The earliest Greek MSS containing the text are of only the fourteenth century; but their similarities show that they were derived from an older Greek archetype that is now lost. However, since we now have an excellent edition of the Greek text by Emilie Boer (1894-1980), *Ptolemaeus III—2 Karpos* (Leipzig: B. G. Teubner 1961. 2nd ed. revised) xvii-xxxiv, 70 pp. I thought it would be useful to have a new translation directly from the Greek.

The Latin version to which I usually refer in the notes is that of Giovanni Pontano (1426-1503), made from one or more Greek MSS that were available to him. It was first published in 1477 and

[123]And it is mentioned by al-Nadîm in his *Fihrist*, ii, 640, which was completed about 990 A.D.

often reprinted by later writers. Since some of the Aphorisms are vague, and in some cases there is a significant difference between the Greek text and Pontano's Latin version, I have occasionally cited the Latin version as well as the translations into English made by Henry Coley (1633-1707) and J. M. Ashmand (fl. 1822). In addition, I have several times cited the earlier Latin translation (as Old Latin), which often differs from Pontano's version. And I have also made some comments of my own.

Where I refer to Greek MSS in the footnotes, the references are to the sigla adopted by Emilie Boer in her edition.

PREFACE TO THE CENTILOQUY

We are setting forth, O Syrus,[124] the effects of the stars, those produced in the complex cosmos and very useful for forecasting, and we have selected the present work, which indeed is Fruit from those books developed through experience. The one intending to pursue this must then first consider all the methods of knowledge, and then come to the reading of this.

1. From you and from knowledge, for it is not possible for the one knowing to proclaim the particular forms of affairs; wherefore, the perception does not receive the particular form of the perceptible thing, but [only] something general. And the one searching must guess at the affairs. For only those who are inspired predict the particular things also.

2. When anyone seeks the better [choice], there will not be any difference between the thing itself and the idea of the matter.[125]

3. The one suitable for a certain affair will certainly also have the star signifying that fortified in his nativity.

4. The mind suitable for foreknowledge obtains the truth more than the one practicing the art the most.[126]

---

[124]Syrus was Ptolemy's patron, to whom his authentic works were addressed in the second century. His mention here is a deliberate attempt by the unknown author of the *Centiloquy* to make his composition look like a genuine work of Ptolemy.

[125]This Aphorism is vague in the Greek. The Latin version is, *Cum is qui consultat, ipsum melius scrutabitur inter id et eius formam, nulla rerum differentia erit.* 'When he who consults, will examine a thing better, between it and its form there will be no difference of things.' This seems to say that a careful astrologer will find no difference between what exists and his astrological view of it.

[126]That is, one who has a natural gift for prophecy will be more successful than one who has mastered a predictive art from books.

5. A skillful person is able to avert many effects of the stars, when he is aware of their nature, and to prepare himself beforehand for the occurrence of the effects.[127]

6. Then the choice of days and hours helps, when it is a convenient time; for, if it is opposite, nothing will be useful, even if it looks at a good outcome.

7. One is not able to understand the mixtures of the stars if he has not first distinguished the differences and the natural mixtures [of things].

8. The sagacious mind helps the heavenly effects, just as the best farmer helps Nature through plowing and clearing [the fields].

9. In generation and corruption, forms are affected by the heavenly forms. And through this the makers of astrological images use them by examining the transits of the stars to them.

10. Use the malefics in the choices of days and hours, just as the best physician [uses] poisons in moderation for therapy.

11. You should not first elect days and hours before you know the maker[128] of the proposed matter.

12. Love and hate hinder making progress toward true astrological decrees. For they make the great things inferior and they magnify the small things.

13. When the heavenly power will denote something, use the helping and the destructive stars, surely the secondary ones.

14. Oh, how many things baffle the learned one, when the 7th house and its ruler are afflicted.

15. The ascendants of enemies of the kingdom are those signs

[127]This Aphorism paraphrases Ptolemy, *Tetrabiblos*, i. 3 "That it [Astrology] is also Beneficial."

[128]The Greek text has *poiotêta* 'maker', while the Latin text has *qualitatem* 'quality', which seems more appropriate.

cadent to its ASC; and the ascendants of the intimates are the angles; and the ascendants of those dwelling in it are the succedents; the same also in dogmas.

16. When the benefics rule the 8th house, they cause hurt from good men, but if they are made fortunate, they release this.

17. When you are forecasting about the life of an old man, do not forecast first before you measure out how many [years] he may live.[129]

18. When the two Lights are in one [and the same] minute, and also a benefic is in the ASC, in everything there will be good fortune in those things the Native [takes] in hand. Similarly too, if the lights are opposing each other from the ASC and the DSC. And if there is a malefic in the ASC, think the opposite.

19. The effect of a purge is impeded when the Moon is conjoined to Jupiter.

20. You should not touch a bodily part with iron when the Moon is present in the sign that rules that part of the body.

21. When the Moon is in Scorpio or in Pisces, and the ruler of the ASC is in aspect to a star that is under the Earth, it is good for purgatives to be employed. And if it is in aspect to a star above the Earth, he will vomit up the drink.

22. Do not don or cut out a garment when the Moon is in Leo. And if it should be impedited, it is still worse.[130]

23. The configuration of the Moon with the stars makes the Native to be moveable; and if the stars are strong, they denote a suc-

[129]That is, his physical condition and life expectancy should be considered before making an astrological judgment.

[130]Henry Coley adds this note to his translation of Aphorism 22: "Mr. Lilly saith, once casually, without inspection of the position of the Moon, he put on a new Suit, the Moon being in Leo, and ill dignified, and tore many holes in the Suit going a Nutting, within a fortnight after; nor did that Suit ever do him any service." Cf. also *The Centiloquy of Hermes Trismegistus*, Aphorism 82.

cessful motion; and if they are weak, a useless motion.

24. The eclipse of the Lights occurring in the angles of a nativity, or in those of the interchanges of the years,[131] is noxious. But take the time from the distance between the ASC and the place of the eclipse, and just as you take the times[132] from the hours of the solar eclipse, so also take the months from the hours of the lunar eclipse.

25. Make the progression of the ruling Planet, when it is in the MC, by the ascension of the right sphere; but when it is in the ASC, by the ascension of the clime.[133]

26. The matter is entirely hidden, when the star signifying it is conjoined to the Sun or is under the Earth or is in an alien place.[134] But the matter is manifest when the star is removed from its fall to its exaltation, and when it is in an angular house.[135]

27. Venus provides joy for the Native in that part [of his body] which the sign, in which she is located, rules. The same thing also [applies] in the case of the rest of the stars.

28. When you are not able to put the Moon to conjoin two stars, put her to join a fixed star having the nature of the two.[136]

29. The fixed stars grant unexpected and marvelous good fortune, but as in most cases they confirm the misfortunes.[137]

---

[131]The translators all take this to mean 'a Revolution of the Years of the World', i.e. an Aries Ingress, or a Revolution of the Years of a Nativity, i.e. a Solar Return. The Old Latin version has *revolutionum annorum* 'Revolutions of Years'.

[132]The Latin text has *annos* 'years' where the Greek text has *chronous* 'times'.

[133]That is, use right ascensions for a Planet in the MC, but oblique ascensions for one in the ASC.

[134]This may simply mean 'not in its own domicile', or perhaps 'when it is peregrine'. The Old Latin has, . . . *vel fuerint sub terra: aut in loco non congruenti domui eorum vel exaltationi* 'or when they are under the Earth, or in a place not agreeing with their domicile or exaltation'.

[135]Old Latin . . . *in angulo in quarta vel septima* 'in the 4th or the 7th angle'.

[136]According to Ptolemy, *Tetrabiblos*, i. 9, most of the fixed stars have the nature of two planets.

[137]The Old Latin has *sed multotiens finisitur in malum* 'but oftentimes ends in

30. See in the proclamations of kings if the ASC of the proclamation harmonizes with the ASC of the begetting of the child of the king, [if so,] that one will be born the successor to the kingdom.

31. When the ruling Planet of the kingdom falls into a climacteric place, the then king or some great archon of it will die; and when a star in the sign of the end finds it in the interchange of the years of the kingdom, something great [occurs] according to the nature of that star.[138]

32. The amicable aspect of the stars in each of the nativities, denoting the kind of the affair by which the friendship is established, makes concord between two persons.

33. Friendship and hatred are taken from the harmonies and disharmonies of the Lights and of the ASC's of each nativity; and the hearing signs[139] increase the friendship.

34. The ruling Planet or the chart ruler of the new Moon[140] being angular denotes all the things that are happening in each month.

35. When the Sun falls into the place of a star, it excites the effect of it on the air.

36. Use the helping fixed stars in the building of cities, and the planets in the building of houses. And the kings of every city having Mars in the MC are killed by the sword in most cases.

37. Those having Virgo or Pisces rising will be the cause of their own authority; and those having Aries or Libra ascending will be the causes of their own death <and Scorpio or Taurus will

---

evil'. The new Latin text adds, *nisi et planetae ad felicitatem conveniant* 'unless the Planets also agree on good fortune', which appears to be a translator's comment.

[138]The Latin version omits the latter part of this Aphorism ('and when a star ... of that star'.).

[139]The *hearing* (or *obedient*) signs are the southern signs from Libra to Pisces.

[140]The Greek has *tou synodou* 'of the conjunction', but the new Moon is meant. Hereafter in this translation, I shall translate it as 'new Moon' without a footnote.

be the cause of sickness>.[141] And consider the rest of the signs in the same manner.

38. When Mercury is in one of the domiciles of Saturn and it is fortified, it gives the Native a sagacious mind and an accurate [understanding] of matters. But in the domicile of Mars, it gives <strength and sharpness of treachery or good> glibness of tongue, and especially in Aries.

39. The affliction of the 11th house at the proclamation of a king denotes damages in his household; and the affliction of the 2nd denotes damage to his flock of sheep.[142]

40. When the ASC is held in check by malefics, the Native delights in ugly actions, and he approves of ill-smelling odors.

41. Be careful of the affliction of the 8th [house] and its ruler at the time of a departure, and of the 2nd and its ruler at the time of an arrival.

42. When a sickness begins with the Moon in the sign in which any one of the malefics was in the nativity, or in its square or opposition, it will be very annoying; and if a malefic is in aspect, it will be dangerous; and if it is in the place, in which there was a benefic in the nativity, it will not be dangerous.

43. When they are configured adversely during their passage, they heighten the malefic configurations of the nation.[143]

---

[141]Adds the Old Latin version.

[142]The majority of the Greek MSS have *blabên tês poimnês* 'damage to his flock of sheep', but MS β has *blabên tôn chrêmatôn tês poimnês* 'damage to the wealth of his flock of sheep'. The Latin version has *pecuniarum ovilis eius detrimentum* 'loss of the wealth of his sheepfold'. Coley and Ashmand take it another way. Coley: 'his Subjects shall be impoverished under his Government. Ashmand: 'the detriment of his subjects' wealth'. Both of them evidently taking 'flock of sheep' to refer in a figurative sense to the king's subjects; rightly so, probably, since the Old Latin has, *...significans quod modicum adipiscetur populus cum illo rege* 'signifying that the people will get a small amount with that king'.

[143]This Aphorism seems to say that adverse transits heighten any evil aspects in a national chart. Ashmond and the Marquis get it nearly right, but Coley evidently didn't understand it, and he cited a lengthy passage from Haly (who had also faield to understand it).

44. When the ASC of a sick person[144] is contrary to the configurations of his nativity, and the time will not bring back something beneficial,[145] it is contrary [for him].

45. Every man not having the ruling Planet of his nativity or the ASC in signs of a human form will be estranged from men.

46. Great good fortune in nativities results from the fixed stars, and both from the angles of the new Moons[146] and from the place of the Lot of Fortune of the kingdom, when the ASC is fortunately [posited] in them.

47. When a malefic in one nativity falls on the place of a benefic in another nativity, the one having the benefic will be damaged by the one having the malefic.[147]

48. When it happens that the MC of the ruler is the ASC of a subject, and the chart rulers are configured in an aspect of friendship, they will remain inseparable for a long time. And it is the same if the 6th house of a servant happens to be the ASC of his master.[148]

49. When the ASC of a subject happens to be the MC in the nativity of his lord, the lord will place his trust in that subject.[149]

[144]He means 'when the ASC of the decumbiture of a sick person is contrary' . . .

[145]Reading *epi tina kalopoion* 'then something beneficial' with MSS βγπ rather than *epekeina* 'yonder' with Boer and MSS λ. The Latin has *ad beneficum* 'for a benefit'.

[146]That is, 'from the <preceding> new Moons'.

[147]The Old Latin adds, *Et genus mali erit ex natura planete. Illud vero in quo fiet: erit ex quantitate nature loci eius*. 'And the kind of evil will be from the nature of the planet. But that in which it will be made, will be from the quantity of the nature of its place.'

[148]Coley adds a note, "it may be thought the late Duke of *Buckingham* had such a one." Presumably he refers to George Villiers (1592-1628), the First Duke of Buckingham NN 607, who was a favorite of King Charles I (1600-1649) NN 614. But if so, Coley must have used a different chart for one or the other of those noblemen, because the sources mentioned give both of them Leo rising. The Old Latin version is different and speaks only of royal ministers and the king.

[149]The MSS of the β family add *tosouton hoti hina prostassêtai hup' autou* 'so much so that he will be commanded by him', which is similar to the Latin *ut ei ab illo imperetur* 'so that he will be commanded by him'. The Old Latin speaks first

50. Do not pass over the 119 conjunctions of the Planets; for in them lies the understanding of the things being generated in the cosmos, of generation and corruption.[150]

51. Where the Moon is at the time of birth, that sign will be in the ASC at conception; and where it is at conception, it will be in the ASC at birth.[151]

52. The ruling Planets of the nativities of tall men are in their apogees, and their ASC's are in the beginnings of the signs; and those of short men are being found in their perigees, and in the long-rising and short-rising signs.[152]

53. The chart rulers of thin men have no latitude, but those of stout men do have latitude; and if the latitude is south they will be agile, but if it is north, they will be sluggish.

54. The chart rulers of a construction joined to stars under the Earth impede the erection of the structure.

55. The damage of Mars to ships is lessened, when it is in neither the MC nor in the 11th house; for in those houses it destroys

---

of ministers and then says, *Et similiter si fuerit .vi. alicuius nativitatis in medio celi: ibi recipiens dispositionem a domino ascendentis eiusdem nativitatis: erit boni dominii in suis servientibus*... 'And similarly if the 6th [house] of any Nativity is in the MC and there receiving ther disposition of the ruler of the ASC of that same Nativity, he will exercise good rulership among those serving him'.

[150]Coley adds this note: "What the conjunctions are, and how to understand them, you may have recourse to Mr. *Lilly's Prophetical Merlin*, fol. 51." The Old Latin lists all 120 of them in seven categories: of two Planets, of three Planets, of four Planets, of five Planets, of six Planets, and finally of seven Planets.

[151]MSS βγπ add *ê to toutou diametron* 'or its opposite'. Boer thinks this may have been added from Porphyry 38 or Hephaestio ii. 1, but the Latin also has ...*aut eius oppositum* . . . 'or its opposite'. The second century astrologer Vettius Valens, *Anthology*, i. 21, has a whole elaborate chapter on this subject. The idea of the interchange of the Moon and ASC signs evidently goes back to the Alexandrian founders of Horoscopic Astrology.

[152]The Latin version substitutes *in sublimitatibus 'in elevations' for en tois apogeiois 'in the apogees' and in humilitatibus 'in humilities' for en tois perigeiois* 'in the perigees'. The Old Latin has, *in summitate suorum circulorum* and *in inferiore parte circulorum suorum* 'in the top of their orbits' and 'in the lower part of their orbits', which would be the apogees and perigees. That version also adds that 'the ASC's of short men will be at the ends of the signs'.

the ship, with pirates taking control of it; and if the ASC is also afflicted by one of the fixed stars that are of the nature of Mars, it will be burned.

56. In the Moon's first quarter <i.e. from the end of its conjunction with the Sun>[153] the humors of bodies flow out until the second [quarter]; in the rest [of the quarters] they decrease.

57. When you see the 7th house and its ruler afflicted in the case of a sick person, change the physician.

58. Look at the place of the conjunction—in which house of the ASC of the Time[154] it is; for the accident will occur, when the profection will arrive there.[155]

59. Do not point out about an absent person that he has died, before you look to see that he is not drunk; and that he is not wounded, before you judge, lest he ever have blood drawn; and do not judge that he finds wealth, before you inquire into that perhaps he has received a deposit; for the configurations of all these may be similar.

60. In the case of sick persons, look at the critical days and the progress of the Moon in a sixteen-sided figure.[156] For when you find those angles not impedited, it will be good for the sick person, but if they are afflicted, the contrary.[157]

61. The Moon signifies those [parts] of the body that resemble

---

[153]Adds βγ; and the Latin version adds, *hoc est ex quo a Solis coniunctione* 'i.e. from its conjunction with the Sun'.

[154]That is, the ASC of the chart for the Revolution of the Years of the World.

[155]J. B. Morin in his *Remarques Astrologiques* (Paris: P. Ménard, 1657; Paris: Retz, 1976. repr.) finds fault with this Aphorism and devotes 5 pages to discussing it, but I don't see anything wrong with it.

[156]Coley points out that Lilly's *Christian Astrology* has a table on p. 294 to assist the astrologer in making his calculations. Morin observes (*Remarques Astrologiques*) that the mention here of a sixteen-sided figure shows the distinction between the author of these Aphorisms and Ptolemy, who only mentions a twelve-sided figure.

[157]The Old Latin's Aphorism is more involved, and Haly has supplied an extensive commentary.

her according to the motion.[158]

62. When you make the minute [of the] beginning of the new Moon,[159] you can predict the change of the air in that month; for the prediction will be according to the ruler of the angle of each figure,[160] for that one controls the nature of the air, taking into account the temperature of the current season.

63. See, when Saturn and Jupiter are in conjunction, which of these is superior,[161] and predict according to the nature of that one. Do the same also with the other twenty[162] conjunctions.

64. In the least conjunction, the difference of the mean conjunction, and in the mean conjunction the difference of the greatest conjunction.[163]

65. When you see the chart ruler of the Question, look at what power that one has in the exchange of the years,[164] or in the ASC of

---

[158]The Old Latin has, *Luna propria est corpori propter consimilitudinem eius in operatione*. 'The Moon is proper to the body because of her similarity in action.' This Aphorism is vague in both versions.

[159]That is, when you have calculated the exact moment of the new Moon.

[160]Here, the phrase 'ruler of the angle' probably means 'ruler of the ASC'.

[161]That is, which of them is the most elevated in the chart.

[162]The word *eikosi* 'twenty' is present in MSS λρ, MS π has *duo* 'two', MSS βγ omit the number, and one of the Latin translations has *XX* 'twenty', but Pontano's version also omits the number, and both Coley and Ashmand omit it. But 'twenty' is correct, for Saturn has 5 other conjunctions, Jupiter has 5, Mars 4, the Sun 3, Venus 2, and Mercury 1, the sum of which is 20. The Old Latin version also has *viginti* 'twenty'.

[163]Of this Aphorism, Ashmand says, "Partridge has said, 'how Ptolemy meant it to be understood, I know not, and so I leave it'." Coley does not comment, but instead prints the whole Aphorism in italics. J. B. Morin, *Remarques Astrologiques* (Paris, 1657), devotes three pages to discussing this Aphorism, but his comments can be summed up by saying that while "this Aphorism has always been believed to be one of the great secrets of astrology," in Morin's opinion "it is a piece of nonsense that teaches nothing and that has only confused" everyone.

The Old Latin adds, *Cum igitur locutus fueris de divisione partium confirma universitatem: et non ponas sermonem tuum in revolutione: quia vertetur malum esse in bonum*. 'Therefore, when you have spoken about the division of parts, confirm the universality; and you should not place your statement on a Revolution because evil is turned into good.' Haly thinks that it refers to a distinction between major influences and minor ones. But if so, it is still vague.

[164]That is, in the Revolution of the Years of the World or the Aries Ingress.

the conjunction,[165] and predict according to them.[166]

66. Do not use profection alone, but also the additions and subtractions of the stars.[167]

67. The additions of the stars[168] are reduced by the weakness of the one receiving them.

68. When the malefic is matutine, it signifies mishaps; but if it is vespertine, it signifies illnesses.

69. When the Moon is opposed to the Sun and is conjunct nebulous stars, damage to the eyes occurs. And when the Moon is in the DSC, and the two malefics are in the rising [part of the chart], and the Sun is angular, he will become blind.

70. In the case of those inspired by a god, the Moon is not configured with Mercury and neither one with the ASC; in the case of the raving madmen, for this configuration Saturn is angular by night, and Mars by day, and especially in Cancer and Virgo and Pisces.[169]

71. In the nativities of men, when the two Lights are in masculine signs, their activities are according to nature; but in those of women such activities are increased; the same thing also [happens] in the case of Mars and in the case of Venus [when they are]

---

[165]That is, the new Moon.

[166]In the Greek text, Aphorisms 64 and 65 are in reverse order from the Latin version.

[167]Boer and MSS βγ have *kai tais lêpsesin* 'and the subtractions', but MSS λπρ have *en tois etesin* 'in the years'. Coley says, "*This Aphorism hath relation unto Nativities, and is for discovery of the true time of the Natives Death.*" It does seem to be related to the next Aphorism.

[168]Boer has supplied *asterôn* 'stars' where MSS MSλπ have nothing, and MSS Dγπ have *etôn* 'years'. Some of the Latin versions also have *anni* 'years'. This Aphorism is vague. Coley explains it thus, 'The years of the Native are deminished and made much shorter, by reason of the imbicilty of the giver of life'.

[169]The Old Latin has a significant difference! It has *epilentici sunt in quorum nativitatibus luna non complectitur mercurio* 'Those in whose nativities the Moon is not configured with Mercury are *epileptics*'. Cf. Ptolemy, *Tetrabiblos*, iii. 14 "Of Diseases of the Soul." where exactly this statement is made.

matutine, for they are being masculine, but when vespertine they are being feminine.

72. Take the [indications] of rearing from the rulers of the triplicity of the ASC, but those of livelihood from the triplicity rulers of the Light of the Sect.[170]

73. When Mars[171] happens to be with the Head of the Gorgon[172] and not aspected by a benefic, nor is there a benefic in the 8th [house], and the ruler of the Light of the Sect is opposing or squaring Mars, the Native will be beheaded,[173] and if the Light is in the MC, his body will also be crucified. But if the application is made from Gemini or from Pisces, his hands and feet will be cut off.

74. Everyone having Mars in the ASC will certainly have a scar on his face.

75. When Mars is in conjunction with the ruler of the ASC in Leo, and Mars is not a chart-ruler [located] in the ASC, nor in the 8th house, the Native will be burned up.[174]

76. When Saturn is in the MC, and the Light of the Sect opposes him and it is in the IMC in an earth sign, the Native will come to an end from the collapse of buildings. And if it is a water sign, the Native will be drowned in water. And if it is a sign of human form, he will be drowned by men, or by strangulation, or by being whipped. And if there is a benefic in the 8th [house], coming close to those,

[170]That is, from the Light of the Time—the Sun by day, and the Moon by night. The Old Latin adds (before 'and the MC") *et gradum solis potestati principum eius* 'and the degree of the Sun for the power of his authority',

[171]Boer following MSS DSλπ has *Arês* ' Mars', but MSS Mγ have *hêlios* 'Sun'. The Latin versions used by Coley and the Marquis de Villennes had 'Sun'; and whatever version used by Ashmand also had 'Sun'.

[172]The star Algol or β Persei.

[173]Like the Gorgon, who was beheaded by the hero Perseus.

[174]Here Boer's Greek text disagrees with the other versions in several respects. First, MSS DSλπρ have Mars aspecting the ruler, while MSS Mγ and the Latin versions have the Sun aspecting the ruler. Second, her edition has '[Mars] is not in the 8th house', while the versions used by the translators have 'no benefic is in the 8th house', which seems more likely to be correct. Her apparatus does not show any such variant.

he will not be slain.

77. Make the profection of the ASC for matters of the body, but the Lot of Fortune for acquisitions, and the Moon for the connection of the body with the mind, and the MC for actions.

78. A star often acts in a place where it has no efficacy, producing unexpected gain for the Native.[175]

79. When Mars is in the 11th [house], the one having it thus does not rule his own master.[176]

80. When Venus is conjunct Saturn and has some rulership in the 7th [house], the *native's* sexual relations will be dirty.[177]

81. Times are reckoned in seven ways. From the distance between two significators; from the distance between their aspects; by the transit of one to the other; from the distance between one of them and the house showing the thing sought; from the giving of the star with its correction[178] and from the progressions and phases and similar things; and from the transits of the stars into its place of domicile.

82. When the configurations are equal, look at the ASC of the new Moon or the full Moon, and if they are equal, do not rush into the astrological significations.

---

[175]Both Coley and the Marquis de Villennes say that this refers to the antiscion of a Planet. But antiscions are not mentioned anywhere else in this *Centiloquy*, nor are they mentioned in the *Tetrabiblos*. The Old Latin makes a contrary statement, *Nil operatur planeta in loco in quo nil promiserit: nec in loco ad quem non pervenit Nativitas.* "A Planet does nothing in a place where it has promised nothing, nor in a place to which the nativity does not come'.

[176]The Old Latin has, . . . *dominus nativitatis infidelis erit principi suo vel regi.* '. . . the ruler of the Nativity will be unfaithful to his own prince or king'.

[177]Ashmand mistranslates the latter part of this Aphorism as 'the native will be of spurious origin'. But the Marquis de Villennes and Coley get it right, and the Old Latin agrees with them.

[178]This way is uncertain. The Greek text has *apo tês doseôs* 'from the giving', but some of the MSS have *apo tês dysis* 'from the DSC', and that is what the Latin has (*ab occasu*), but the latter does not make sense. Possibly the intent was merely to derive a time by using the "corrected" place of the Planet rather than its mean place.

83. The time of putting forward shows the arrangement between the one being put forward and the king; and the chair[179] shows the [results] of the action.

84. When Mars is the ruler of the ASC at the time of taking one's seat, and it is in the 2nd [house] and joined to the ruler of the 2nd, it will cause much damage.[180]. When the ruler of the ASC is configured with the ruler of the 2nd [house],[181] the leader will willingly make many expeditions.

86. The Sun is the source of vital power; and the Moon of natural power; and Jupiter of growth; and Mars of boldness; and Mercury of reasoning; and Venus of desire.[182]

87. The exchanges[183] of the months are of 28 days and 2 hours and 18 minutes very nearly.[184] But some judge them from the position of the Sun, when it becomes equal to the degree that it had at the beginning.

88. When we want to make the progression of the Lot of For-

---

[179]The word 'chair' by itself does not seem to make any sense. Perhaps we should read <*ho de kairos*> **prokathedrias* '<and the time> *of sitting in the chair'. (The Old Latin has, *hora vero intronizationis eius* 'but the hour of his enthronement'.) This would refer to a Horary chart set for the time certain of a person's taking a seat of authority or prestige. Cf. Aphorism 84. And see Coley's remark, "the time when the Petitioner receives the dignity granted him, shall shew the quality of action...."

[180]The Old Latin adds, 'especially if Jupiter is the ruler of the house'.

[181]The old Latin continues at this point with, 'the steward or manager will consume or expend much. But if it is debilitated, he will lose much. And when the ruler of the 2nd commits its disposition to the ruler of the ASC, he will prosper', etc..

[182]The translations of the Marquis de Villennes, Coley, and Ashmand omit the words from 'and Jupiter' to the end of the sentence, since they are not present in the Latin version.. Boer's apparatus shows that some of the Greek MSS also omit them

[183]That is, the Revolutions.

[184]According to modern figures, the length of the synodic month is 29 days 12 hours and 44 minutes, and the length of the tropical month is 27 days 7 hours and 43 minutes. But J. B. Morin, commenting on this Aphorism, explains Hermes's figures. They are derived by dividing the tropical year of 365 days 5 hours and 49 minutes by 13, which does in fact give Hermes's figure of 28 days 2 hours and 18 minutes. This is the length of a "profectional month" that was used by some of the old astrologers.

tune for the whole year of the Revolution, we take [the distance] from the Sun to the Moon, and [we cast] an equal amount from the ASC.

89. See the [affairs] of the grandfather from the 7th house, and those of the uncle from the 6th house.

90. When the ruler aspects the ASC, the thing that is concealed will be of the nature of the ASC; and if it does not aspect it, it will be of the nature of the place where the ruler is; and the Ruler of the Hour shows its color; and the house of the Moon shows the time, for if it is above the Earth it will be new, but if it is below the Earth it will be old. And the Lot of Fortune shows the size of it, if long or short; and the ruler of the terms of the 4th [house] and the MC and of the Moon shows its nature.

91. When the ruler of a sick person is under the Earth,[185] it is a bad sign, and especially if the Lot of Fortune is afflicted.

92. Saturn being oriental does not afflict the [condition] of a sick person so much, just as Mars does not when he is occidental.

93. You should not predict from the configurations first before you look carefully at the new Moon, for the beginnings are changed in every new Moon; therefore, mix both of them together, and you will not err.[186]

94. The house of the most potent of the significators shows the things that are in the Querent's object.[187]

95. It is likely <that>[188] the *paranatellonta* of each decan[189]

---

[185]The old Latin has, 'under the Sun beams'.

[186]Coley has "... the next subsequent new Moon . . ." Ashmand has "...until the next conjunction ... and ...both the last and the next [conjunctions] should be combined." But neither Pontano's Latin version, nor the Greek text support these extra words.

[187]This is a rule for what is usually called "the divination of thoughts," or what the Querent had in mind that motivated him to ask a question.

[188]The other translations omit this phrase.

[189]These are the figures that are said to rise with each decan. There was an exten-

show the *native's* preference and the art that he practices.

96. The astrological indications of an eclipse are [from] those nearest the angles. And also look at the natures of the conjoining stars, the Planets, and the fixed stars, and the *paranatellonta*, and from them give an explanation.

97. When the ruler of the new Moon or of the full Moon is angular, the matter about which there is a question is [soon] brought to completion .

98. {Consider the ruler of the angle of a new Moon and of a full Moon, as well as those of the quarters of the year. When it is strong, that which it signifies will be the substance [of that time period], and [see] if it is in angles or in succedents. And when it is weak, it will be bad, namely when it is in cadent houses; and similarly [see] when it is swift or slow.}[190]

99. [191]The shooting stars and comets have the second place in astrological judgments.

100. [192]The shooting stars show the dryness of the air; and if

sive literature on these in the middle ages. Presumably the figures were originally related to the constellations (both Greek and Barbaric) that rose with the individual decans. Here is an example from Ibn Ezra, *The Beginning of Wisdom*, Chapter 2, "In the first face [of Aries] there will ascend a figure of a woman, which is the radiant one, the tail of the sea-fish resembling a serpent, the head of the triangle, and the form of an ox," and he also gives the opinions of the Hindus and Ptolemy.

[190]There is no Aphorism numbered 98 in the Greek MSS, so Boer has furnished a Latin text, which I have translated above in braces. However, the Latin text used by Coley, the Marquis de Villennes, and Ashmand was entirely different. The Latin text is this: *Trajectiones atque crinitae, secundas partes in judiciis ferunt*, which Ashmand translates aptly as "Shooting stars and meteors like flowing hair bear a secondary part in judgments." And this is in fact the Aphorism numbered 99 in the Greek text.

It looks as if the scribe of the lost Greek archetype copied Aphorism 97 and then Aphorism 98, but he mistakenly numbered it Aphorism 99. I have put a virgule in his Aphorism 100 to mark the end of what should have been numbered Aphorism 99 and the beginning of Aphorism 100. The situation is the same in the Old Latin version.

[191]This corresponds to Aphorism 98 in the Latin version and the translations derived from it.

[192]This is misnumbered in the Greek version. It is actually Aphorism 99 down to

they bear into a particular part, they show the direction of the wind from that origin; but if they bear into different parts, they show a lessening of water and a turbulence of the air and incursions of soldiers. [99|100] And the comets, whose stand is in the eleventh sign from the Sun, if they appear in an angle of some kingdom, the king of it or some great man of it will die But if they appear in a succedent house, the things of the treasuries of it will have good, and it will change the government of it. But if they appear in the cadents, they produce sick people and sudden deaths And if they move from the west to the east, some other enemy will attack those regions; but if it does not move, the enemy will be indigenous.

---

the virgule, and Aphorism 100 thereafter. See the Note to the end of Aphorism 98 above.

THE CENTILOQUY OF HERMES TRISMEGISTUS

TRANSLATOR'S PREFACE

This *Centiloquy* was translated from Arabic sources by Stephen of Messina and addressed to Manfred (1231-1266), King of Sicily, perhaps in the year 1262. It is in good Latin, and it was very popular. Many other astrological tracts of various sorts attributed to Hermes appeared in Arabic in the Middle Ages and were translated into Latin in the 12th century; their actual origin and authors are unknown. This *Centiloquy* was first printed by Erhard Ratdolt at Venice in 1484 in an omnibus edition of astrological works. There were many subsequent printings. I have used the one printed in an omnibus edition of astrological works at Basel in 1533 by Ioannis Hervagius.

I believe the first English translation of the Latin was made by Henry Coley (1633-1707) and included in his book, *Clavis Astrologiae Elimata* (London: Ben Tooke and Thomas Sawbridge, 1676), pp. 329-339. It has since been reprinted several times. However, I thought that a new translation might be appropriate, since I am repeating Coley's presentation of all three *Centiloquies* in a single volume.

THE CENTILOQUY

1. The Sun and the Moon, after God, are the life of all living things. The nativities of many persons do not have a Hyleg,[193] but because the Sun and the Moon aspect their ASC from a friendly spot, and are at the same time free [from the malefics], their lives are prolonged for a long time.

2. All diurnal nativities are strengthened by the Sun when it is configured with the benefics. But nocturnal nativities are strengthened by the Moon when it is aspected by the benefics. And if this is not the case, but nevertheless there are good planets in the angles, the Nativity is thus made fortunate.

3. When Mars, ruler of the ASC, is in the 10th, it gives the Native dignity and power, which he will use with injury and cruelty; and this is truly said to be a misfortune rather than good fortune.

4. Jupiter configured with the malefics changes their evil into good. Venus cannot do this unless she is aided by Jupiter; and therefore, in increasing good and in warding off evil, Jupiter is considered to be much better than Venus.

5. An astrologer cannot make a combination of the significations of the stars before he knows their friendships and enmities, of which there are three kinds—one indeed according to their nature, another according to their domiciles, and the third according to their aspects.

[193]Hyleg is from the Arabic word *al-hîlâj* from the Persian word *hîlâk* 'letting loose' that was the equivalent of the classical Greek word *aphetês* 'the aphetic place (lit. 'starter'). It was the significator of the life of the Native according to Ptolemy, *Tetrabiblos*, iii. 10.

6. Venus is the opposite of Mercury, [who rules] speech and learning, but she embraces pleasures and delights. Similarly also, Jupiter is the opposite of Mars, the former indeed wishes for mercy and justice, but latter impiety and cruelty.

7. You should make the Sun, or any one of the superior Planets, to be the significator of great princes, but the inferior Planets,[194] and especially the Moon, to be the significators of scribes and rustics.

8. An aspect cannot diminish the signification of a conjunction. But a conjunction diminishes the signification of an aspect, for a conjunction is stronger than an aspect.[195]

9. You should not define or elect anything with Scorpio in the ASC, nor when the angles are oblique, or if Mars is in them, for the definition will turn out to be false, and especially because Scorpio is the sign of falsity.

10. A benefic produces evils from the 6th or from the 12th house, when it is impedited there by a malefic.

11. Rumors that are said when the Moon is in the first face of Scorpio[196] are lies and made up.

12. It happens that whenever the judgments of astrologers on Questions, is not true [the cause is] either on account of errors in their instruments,[197] or on account of the foolishness of the Querent, or when the Sun is around the MC degree, or when the figures that allow the thing, or deny it, are equal.

13. When the Moon will be in the south, descending in Scorpio

---

[194]The superior Planets are Saturn, Jupiter, and Mars. The inferior Planets are Venus and Mercury. These terms refer to their position with respect to the Earth.

[195]A conjunction, properly speaking, is a *position*, while the sextile, square, trine, and opposition are *aspects*.

[196]That is, 'in the first *decan* of Scorpio'.

[197]Hermes is referring to the instruments (sundials, or astrolabes) used to determine the exact time when the question was asked. Today it is easy to know the exact time, but it was difficult in the old days before the invention of clocks.

or Pisces, you should not begin to build, for a building of that sort will be demolished.

14. Mercury should be strong through [being in] suitable places and through [having] configurations with other stars in nativities for that which the Native has dignity in, for Mercury by itself is a weak Planet.

15. In bicorporeal signs, victories are good, but defeats are evil, since both of them are doubled [in strength].

16. You should not define anything before you know the intention of the Querent. Many indeed do not know what to ask, nor are they able to express what they are interested in.

17. When you have been asked about a father, look at the 4th, about a brother the 3rd, about a child the 5th, about a wife the 7th. But if you have been asked about a sick person, you should look at nothing else but the ASC.

18. When the Moon has come to the squares of the benefics or the malefics, both impediment and aid will be doubtful; and it must be feared that the malefics will indeed strongly impedite, and the benefics may not be able to aid.

19. In the beginning of journeys and returns [from them], do not let the Moon be in the ASC, nor in the 4th or the 9th, even though she is not impedited. But in going into a city, do not let her be in the ASC, nor in the 2nd or the 4th. Similarly, not in going into a house.

20. There are three ways in which the accidents of men are known—either from their own nativities, or from the birth of their first-born child, or from a Question put with concern and anxiety.

21. Every beginning made when the Moon is conjoined to a retrograde Planet, is quickly destroyed, and it falls into an even worse condition when it is impedited.

22. The condition of kings and princes is had from Saturn and the Sun, and from the Planet that is in the 10th. But the condition of

the aides of the king is had from the 11th, and the condition of those aiding the rustics from the 2nd house.

23. The departure of the king or a prince on journeys is absolutely disapproved when Cancer will be in the ASC.

24. Gemini and Sagittarius obey the Head and Tail of the Dragon[198] more than the other signs; and they therefore make graver evils in them than they do in other signs.

25. In the nativities of women, when the ASC is in any of the domiciles of Venus with Mars in them[199] or when the ASC is in any of the domiciles of Mars with Venus in them, the woman will be shameless; and it will be the same if she has Capricorn in the ASC.[200]

26. The Sun receives the virtues of the Planets when it is in the ASC or in the house of the MC when it is joined to them. The Moon does this similarly by night when it is joined to the aforesaid bodies in the same places.

27. Jupiter dissolves the malice of Saturn, and in the same way Venus dissolves the malice of Mars.

28. When the Question is about a woman, simply accept the signification from Venus, but more particularly from the 7th house. And if it is about an enemy, absolutely indeed from the 12th, but particularly from the 7th.[201]

29. In [a question about] anyone going out to war, but especially about kings going out to war, let the ASC be in any domicile

---

[198]The Head and Tail of the Dragon are the North and South Nodes of the Moon's orbit. The old astrologers assigned them to the signs Gemini and Sagittarius respectively.

[199]In Chaucer's *Canterbury Tales*, we read of the Wife of Bath, who said, "Myn ascendant was Taur, and Mars thereinne. Allas! allas! that evere love was synne!"

[200]A 17th century example was Nell Gwyn (1650-1687), mistress of King Charles II. She had Capricorn rising. See NN 203.

[201]The 12th house is the House of Secret Enemies, while the 7th house is the house of Open Enemies.

of the superior Planets or the Sun, and let its ruler be strong, but let the ruler of the 7th be weak and impedited by the malefics.

30. The Moon joined to Saturn or Jupiter, when she is increased in light and number,[202] it will be good for all things. But if she has little light, it will be evil for all things.[203]

31. Take care in borrowing and lending that Jupiter not be under the Sun beams[204] or impedited by the malefics. But if it is so, and it is not received by the source of the impediting, no or only modest restitution will follow.

32. When the benefics are going to configurations with the malefics in any figure, they diminish some of their evil; and in good figures, they do more [in this way], but in bad figures, less. But when the malefics are going to configurations with the benefics, namely by square or opposition, they diminish something of their good, but by other aspects they do nothing.

33. When Saturn is transiting from one sign to another, there are made in the sky shooting stars,[205] that the Arabs call *Assub*,[206] or some other [celestial] signs of the nature of fire.

34. Temperate air results from a conjunction of Jupiter and the Sun, especially when that conjunction is made in air signs. Cold is made from a conjunction of Saturn and the Sun. And from a conjunction of the Sun and Mars in a bicorporeal sign and in the season of spring, there is made a darkness of the air; and illnesses frequently occur.

---

[202]'Increased in number' means 'swift'.

[203]Coley adds, "understand the contrary wholly, when she is in conjunction of Venus and Mars."

[204]That is, *combust*.

[205]The Latin text has the word *diachohontes*, which is a corruption of the Greek word *diaittontes* 'shooting stars'. This is very interesting, because it might indicate that the Latin text was translated from a Greek original or that the translator had some knowledge of Greek.

[206]This is the Arabic word *ashhub*, an old plural of *shihab* 'shooting star' or 'meteor'. It is mentioned by Albumasar in his book *De magnis conjunctionibus* 'The Great Conjunctions'.

35. In the season of summer, when the Sun enters the terms of Mars, heat is caused; in the winter, there will be dryness and a scarcity of rains.

36. In the nativities of men and in Questions, you should make the Hyleg and the *Alcochoden*[207] and their directions, and especially in the case of questions about kings and magnates. For through these their accidents are known, whether they are good or evil accidents.

37. When the ASC is good and its ruler is a malefic, it indeed indicates health of body, but mental anxiety and sorrow. But if it is the other way around, you should declare the opposite.

38. You will always note the configurations of the stars, not by their signs, but by their rays.[208]

39. Use the Moon in curing the eyes when it is increased in light and free from any aspect of the malefics.

40. If the Part of Fortune is with a malefic in the 4th or the 9th or the 10th, removed from the benefics, it indicates death for a sick person.

41. The good and evil will be lasting when the significator is stationary and in an angle, but readily changeable when the significator is retrograde and in houses cadent from the angles.

42. The ruler of the 2nd has the same force for impediting as the ruler of the 8th. Similarly, the ruler of the 6th and the ruler of the 12th.[209]

---

[207]This is from the Arabic word *al-kadkhudâh*, which is derived from a Persian word meaning 'house lord', which was a translation of the classical Greek word *oikodespotês* 'ruler' (lit. house-lord).

[208]That is, the orbs of the Planets should be taken into account when reckoning the aspects. This is the standard practice of modern astrologers, but the ancients considered aspects to be a function of the signs. For example, any Planet in Aries was considered to be in square to any Planet in Cancer (or Capricorn) regardless of the degrees.

[209]This is the result of the ancient rule that opposite houses share each other's significations. Both the 8th house and the 12th house are evil houses, so their opposites

43. When Mars is occidental in Cancer, not aspecting Saturn or Jupiter or Venus or the Sun, he will be a phlebotomist. But in the same way, when Mars is in Capricorn, he will be a corrupter of men, and a lover of the outpouring of blood.

44. A skilled astrologer happens to err when he is ignorant about the significator that he chooses.[210]

45. Saturn elevated above Venus and square to her makes those who are born shameless and impatient with women. But if Venus is elevated above Saturn, it will make those who are good-natured and inclined towards women.

46. If in anyone's Nativity Mercury is in the ASC oriental and swift, then the Native is eloquent and learned in the liberal sciences.The same will also be the case, if it is in Sagittarius in its own terms in such a condition.[211]

47. The first of the angles is the ASC, the second is the MC, the third is the DSC, and the fourth is the Angle of Earth.[212] But the first of the rest of the houses is the 11th, then the 2nd, afterward the 5th, then the 9th, afterward the 3rd, and after that the 8th. But the 6th and the 12th are the worst.

48. Unless a benefic Planet prevents it, whatever Mars gives, is [not] revoked.

49. You will adapt the Planets to [the quality of] the person from whom you will ask a favor.[213]

50. When the ASC or a Planet is found in the 30th degree of a sign, its signification is in the following sign; but if it is in the 29th degree, it s influence will be in the sign itself; for the virtue of a

---

the 2nd and the 6th houses are also evil in some instances.

[210]Coley has, 'when he mistakes a true Significator for a false one', but I think Hermes meant what I have written.

[211]The terms of Mercury in Sagittarius are 18-21,

[212]The classical name of the 4th House or the IMC.

[213]This is a rule for an Election. Coley paraphrases, "Let your significator agree with his, whom you intend to supplicate."

planet is considered to be in three degrees—namely, in the one in which it is, and in the preceding degree, and the one following.[214]

51. Future things ought to be considered from the conjunctions of the Planets, but present or past things from their separations.[215]

52. If Jupiter is in Cancer, removed from the ASC, and not impedited by anything, the Native will certain be rational and very steeped in knowledge; he will, however, delight in a solitary life, and he will not receive any praise for his own knowledge.

53. There will be many misfortunes in the World when there will be eclipses of both of the Light in a single month,[216] and especially in those locations in which there is a particular signification.

54. When the Moon is in the *Via Combusta*[217] at the time of the beginning of a journey, the traveler will get sick on his journey, or he will experience some serious troubles.

55. The time ought to be considered in directions of the Planets, but not in directions of the Fixed Stars.[218]

56. The father's estate passes to the son when Saturn is fortunately placed and in a friendly aspect to the ASC, but more so and more fully if it is the ruler of the 4th house.

---

[214]This Aphorism has confused some readers. When the old astrologers spoke of "the degree" of a Planet resulting from calculation, they simply ignored the minutes and only mentioned the degree number. Hence, a Planet in say 30°45 would be said to be in "the 30th degree of Aries," but since a sign has only 30°, it was plainly in the following sign, Taurus; and that is what Hermes is saying in the first part of the Aphorism. But in the last part he is saying that if you only consider the degree *numbers*, then a Planet in the "29th degree" stays in its sign because a Planet in say 28°45 , 29°45 , or 30°45 would be said casually to be in the 28th, 29th, and 30th degree of Aries. And so, a Planet in the 29th degree (29°45) of Aries would remain in Aries.

[215]And by "conjunctions," he means "applications going to conjunctions," because these are configurations that are going to conjunction in the future.

[216]This can happen from time to time, for example in 2005 there was an eclipse of the Sun on the 8th of April and an eclipse of the Moon on the 24th of April. The Sun was eclipsed in Aries, but the Moon was eclipsed in Scorpio.

[217]A hazardous place in the zodiac between 15 Libra and 15 Scorpio.

[218]I don't know what he means by this, unless he is referring to the fact that the Planets move in a short period of time, while the Fixed Stars do not.

57. When the benefics are posited in signs in which they have no dignity, their goodness will be reduced.

58. If Mars is the *Almuten*[219] in a Nativity, and it is not joined to benefics, it signifies that the Native will experience burning.

59. A benefic Planet gives immense happiness when it is received in its own domicile; but a malefic will abstain from much evil when it is similarly received.[220]

60. When Mars is elevated above Saturn, the Native will be sickly and weak. but is Saturn is elevated above Mars, he will be strong and fat.[221]

61. If the Part of Marriage of a Man[222] falls in an obedient sign, but a woman's in a commanding sign,[223] the woman will indeed rule the man, and the man will obey the woman; but if it will be the other way around, you should declare the opposite.

62. If the rulers of the triplicity of the conjunction of the Lights are conjoined favorably in turn—the first to the second, and the second to the third—they signify permanent prosperity and freedom from sorrows.

63. When Mercury will be in the degrees of the *Pits* in Pisces,[224] it makes the Natives to be senseless and mutes. But Jupiter in the

---

[219] *Almuten* is a corruption of the Arabic word *al-mubtazz*, a derivative of the verb *bazza* 'to triumph or to be victorious over someone'. So *al-mubtazz* is the equivalent of the classical Greek word *epikratêtôr* 'ruler', used in the sense of the principal ruler.

[220] This Aphorism does not make much sense as it stands. For when a Planet is in its own domicile, it cannot be received by another Planet (by domicile). Some words would seem to be missing.. Coley has, '. . . when they shall be received <of each other> in their proper Houses . . . '

[221] Coley reverses the Planets in his translation.

[222] This Part is found by the formula, Venus - Saturn + ASC degree.

[223] The northern signs (Aries–Virgo) are usually said to be "commanding," and the southern signs "obeying."

[224] The so-called *Pits* are in degrees 4, 9, 24, 27, and 28 in Pisces. See the table in Lilly's *Christian Astrology*, p. 116, and his discussion of the pitted degrees on p. 118.

domiciles of Mars and in the degrees of the Pits,[225] it makes them sordid and needy and injured by military persons. But in the domiciles of Saturn, and especially in Capricorn, in the degrees of the Pits,[226] it makes inflexible persons and those hated by everyone.

64. Mercury receives the nature of Mars when it is in its domiciles; and if it is configured to Mars in a house cadent from an angle, the Native will be a lover of hunting, and he will love to play dice; but if it is not cadent, he will be warlike and a soldier.

65. Planets under the Sun beams, when they are less than 12 degrees from the Sun, are made unfortunate, unless they are in the same degree with the Sun. but when they go out of the 12 degrees, and are oriental, they are strong.

66. *Caput* with the malefics produces terrible evils, for it will increase their malice. But it produces many good things with the benefics, for their goodness is augmented by it. But when *Cauda* is located in just the same way, it reverses the significations of *Caput*.[227]

67. If anyone has Mercury in the 6th house of his Nativity, he will be converted from his faith to another one. And the one who has his Part of Happiness impedited, he will not be steadfast in his own faith.

68. The first sign has rulership in significations when the signification of the thing is made by two signs.

69. The beginning of everything is taken from the Moon, but the end of everything is taken from the ruler of her sign.

70. If in the Revolution of the Years of the World Jupiter is in its own sign or exaltation, oriental, in an angle, and free from the

---

[225]The Pits in Aries are in degrees 6, 11, 16, 23, and 29; and in Scorpio they are in degrees 9, 10, 22, 23, and 27.

226 The Pits in Aquarius are in degrees 1, 12, 17, 22, 24, and 29; and in Capricorn they are in degrees 7, 17, 22, 24, and 29.

[227]*Caput* and *Cauda* are the *Head* and *Tail of the Dragon*; that is to say, the Moon's North Node and its South Node.

malefics, it signifies an abundant harvest.[228]

71. In the case of sick persons, it should be feared when the Moon and the ruler of the ASC are impedited by the ruler of the 8th house.

72. You should not in any way begin a quarrel, a controversy, or a lawsuit when the Moon is ill disposed; for if it occurs in such a condition, you will without doubt be overcome.

73. Every rebellion beginning to be staged at the beginning of a year will succeed without any difficulty.

74. When the Moon is in ruminating signs,[229] or conjoined to a retrograde Planet, it is not good to take a purgative, for it will cause vomiting or some other injuries.

75. Oriental Planets signifying good or evil things, act quickly, but occidental Planets act slowly.

76. The degree of the conjunction of the Lights is in the middle motion of an eclipse.

77. There will be many difficulties and conflicts [indicated] in Revolutions of the Years of the World,[230] when Jupiter and Saturn are in their own exaltations.

78. Be suspicious and careful when a benefic is with a malefic; and you should not be very confident that the malice of the malefic can be entirely averted.

79. Twelve signs originate from the ASC; and the ASC in fact signifies the condition of the body, but its ruler signifies the condition of the mind. You should, therefore, be careful in every way that the ASC and its ruler are not impedited.

---

[228]Reading *plenum annone* 'plenty of harvest' instead of *penuriam annone* 'penury of harvest', since Jupiter is well placed. Coley agrees, saying, "*my Author says* penury, *if time and ill handling have not abused him*."

[229]The ruminating signs are Aries, Taurus, the latter half of Sagittarius, and Capricorn.

[230]That is, in Aries Ingresses.

80. When the planets are in the fixed signs, they signify something lasting; but in bicorporeal signs, something doubtful; and in mobile signs, something convertible either to good or to evil.

81. In secret matters, let the Moon be coming out from under the Sun beams, not going under them.

82. To cut out or put on new clothes when the Moon is in a fixed sign, and especially in Leo,[231] is horrible and dangerous—also, when she is in conjunction or opposition to the Sun, or impedited by the malefics.[232]

83. The Moon has a great power in Questions, unless the ASC is Leo or Sagittarius or Aquarius. For then some of them take away her signification, and especially when the ASC's are in Leo or Aquarius.

84. Saturn is still under the Sun beams when the distance between it and the Sun is less than 15 degrees. Understand the same thing about Jupiter.

85. The Moon in Cancer should absolutely be refused [for the time of] marriages. Similarly, the Moon in Virgo, except in the marriages of spinsters.[233]

86. When a malefic will be oriental in its own domicile or exaltation, it is better than a retrograde and impedited benefic.

87. There will be an impediment in that part of the body that is signified by the sign that was impedited at the time of the Nativity.

88. When the rulers of the triplicity of the Light of the Time[234]

---

[231]Cf. *Ptolemy's Centiloquy*, Aphorism 22.

[232]In the old days, people did not change their clothes every day as they do now; rather, they would make or don new clothes and wear them for an extended period of time. It therefore marked the beginning of a time period in a person's life. And that is the reason for Hermes's cautionary comment.

[233]The Latin has *viduarum* 'of widows' or 'of spinsters'. I think Hermes means 'spinsters' here. But Coley chooses 'widows'.

[234]The Latin has *luminaris, cuius fuerit auctoritas*, literally 'of the Light whose authority it was'. I take this to mean the 'Light of the Time'. Coley, however,

are in angles or succedents of the angles, in their proper places, and removed from aspects by the malefics, then immense good fortune is produced. And if the ruler of the ASC is in good condition, the good will be more and greater.

89. The Planets do the same thing by trine as they do by sextile, but the sextile aspect produces less good or evil than the trine.

90. Saturn produces evil with slowness, but Mars produces it suddenly; and therefore Mars is reputed to be worse in harming.

91. It is said to be a Grand Conjunction when the three superior Planets are conjoined in one of the royal signs; and then they make most powerful kingdoms when they are aspected by the Sun.

92. A hidden word that is sought is immediately unveiled, when the Moon and the Planet to which she applies are in signs having voice, and in the 5th or the 3rd, or in the houses opposite them.[235]

93. A malefic in the 8th will increase its malice, but a benefic there will exhibit neither good nor evil.

94. Neither good nor evil will be perfected, except when the benefic or malefic Planets in a Nativity or in a Revolution aspect the Moon by square.

95. If Mercury is impedited in the 6th, the Native will die in prison. If Saturn is in the 12th and Venus in the 8th, he will suffer a fall.

96. It must be feared if at the onset of an illness the Sun by day or the Moon by night is impedited.

97. The significations of the stars are always changed when in their mutual configurations with each other their latitudes are changed.

---

translates the phrase simply as, 'luminaries'.

[235]Coley has, '...in the first or third Houses...'

[236]Coley renders the last sentence as, "...the cause of the question is confirmed." But I don't think that is correct.

98. The Moon in the 4th, or in the 7th, or in the 9th, or in the 12th indicates the cause of that which has already been asked about. It indicates the same thing if it is separated from Mercury. But if a bicorporeal sign is ascending, and the Moon is also in a bicorporeal sign, that same cause may be asked about again.[236]

99. If a malefic is in its own domicile or exaltation, it will produce a good end, although with slowness. But if the malefic is impedited in the ASC, even though it is in its own domicile or exaltation, it nevertheless produces an impediment and a bad end.

100. The end of the results of every inception is terminated, and of everything that is doubtful about it, by these significators—namely, by the 4th house, and its ruler, and by a Planet that is strong in it. Also, by the Light of the Time and its ruler, and by the Planet to which that Light is conjoined, and by its ruler.

Use these and the others that I have related to you, and with reasonableness you will never err, with the aid of God.[237]

---

[237]Coley omits this farewell statement by Hermes.

THE CENTILOQUY OF BETHEN

TRANSLATOR'S PREFACE

It is supposed that this *Centiloquy* is not by an otherwise unknown Arabic writer, but that it goes back to a work by the Jewish scholar Abraham Ibn Ezra (1092/3-1167) written in Hebrew in 1148.[238] Its title (in English) was *The Book of Consultations of the Stars*, and it was translated from Hebrew into Old French by Hagin le Juif in 1273. It was subsequently translated from the Old French into Latin. And it is a Latin version that was mistakenly attributed to "Bethen" (or "Bethem").

It was first printed at Venice by Erhard Ratdolt in 1484. There were many subsequent printings. I have used the one published at Basel by Ioannis Hervagius in 1533 (pp. 89-93). The Latin text is of poor quality and rather terse. Some words are used in unusual senses, and a few are of uncertain meaning. A literal translation would be jerky and hard to understand. I have therefore resorted to adding enough words to make decent sentences, and I have also used paraphrase more than I ordinarily do. But there are some places where the text does not seem to make any sense.

I believe the first English translation was made by Henry Coley and included in his book, *Clavis Astrologiae Elimata* (London: Ben Tooke and Thomas Sawbridge, 1676), pp. 339-345. Coley appears also to have had trouble in understanding and translating the Latin, and he often abbreviates the longer Aphorisms. I have mentioned some instances of this in a footnote, but not all of them. And he occasionally writes something that is only similar to what the

---

[238]See *The Beginning of Wisdom* (Baltimore: The Johns Hopkins Press, 1939) and *Arabic Astronomical and Astrological Sciences in Latin Translation* (Berkeley and Los Angeles: University of California Press, 1956).

Latin says or is even completely different.

Coley's translation has been reprinted several times. However, I thought a new translation might be appropriate, since I am repeating Coley's presentation of all three *Centiloquies* in a single volume.

THE CENTILOQUY

1. I shall begin the book about the consultations [that are customary] in the judgments of the stars.

2. You should know that when Planets are retrograde, they are like a man who is weak, stunned, and worried.

3. But a cadent Planet is like a man who is dead and has no motion.

4. A combust Planet is like a man who is a captive, in prison, condemned and without power.

5. A Planet static retrograde is like a healthy man from whom health is receding, but some little of it remains.

6. A Planet that is static direct is like a man who is weak, but afterwards he grows stronger, and he will see the beginning of [a return of] his health.

7. A Planet that is besieged is like a man who is fearful between two enemies, that is when it is between the two malefics.[239]

8. When a Planet is between two benefics, it is like a man who is eating, drinking, and free from all evil and fear.

9. When the malefics aspect a Planet from the 4th house, or when it aspects them, it is like a man for whom his own death is al-

---

[239]I have replaced the old terms, *infortunes* and *fortunes* with *malefics* and *benefics*, which refer usually to the planets Saturn and Mars and Jupiter and Venus.

[240]Reading *decurtatio* 'mutilation' rather than *decoratio* 'decoration'. Coley omits the last words.

ready at hand and punishment and mutilation.[240]

10. When a Planet is in aspect with its enemy, it is like a man fearing his enemy.

11. When a Planet is with its own enemy, it is like a man who is fighting with his own enemy.

12. When a Planet is with a Planet friendly and similar to itself, it is like a man to whom hospitality and affection are offered.

13. When a Planet is in the domicile of its partner, it is like a man who is with his partner.[241]

14. When a Planet is cadent from its own domicile or from the sign of its own exaltation, it is like a man who is absent from his own house or from his own city.

15. When a Planet is in its own domicile or in the sign of its own exaltation, it is like a man who is in his own house; and his decree and his own will is carried out – what he wishes to be done is done.

16. But when a Planet is retrograde in its own domicile or in the sign of its own exaltation, it is like a man who is sick in his own house.

17. When a planet is combust in its own domicile or in the sign of its exaltation, it is like a man who has been incarcerated in his own house[242] by his own lord or by his own king.

18. When a Planet is cadent, it is like a man who is troubled and timid.

19. When a benefic Planet is retrograde, it is like a Planet that is made malefic; and when a benefic Planet is cadent from the angles or from the signs of good fortune, it is like a man who is hoping for good fortune and not finding it.

---

[241]Coley takes *socius* 'partner' to mean a Planet that is a joint ruler of the triplicity.

[242]That is, he has been placed under house-arrest.

20. When a benefic Planet is retrograde and it is with a malefic, it is converted from a benefic to a malefic, and its malice is strengthened.

21. When a Planet that is a malefic is direct in its own domicile, and it is with a fortunate Planet, it is converted from a malefic to a fortune.

22. When a Planet is in the last part of a sign, it is falling and departing.

23. When a planet is in the first degree of a sign, it is of weak virtue in judgments and questions.

24. From the 1st degree to the 15th it is approaching; but from the 15th to the 25th it is completed.

25. If a Planet is in the last part of a sign, it is departing; and then it is like a man who is departing from the house in which he has stayed.

26. A Planet that is feral in the ASC, in which it has authority, is singular in its judgment and action.

27. A Planet that is not in its own domicile, is like a man striking out in another man's house; and he is already deprived of his own judgment, and he has no authority.

28. When a Planet is with the Sun, its powers are weak, and its judgment is diminished.

29. When a Planet is retrograde while it is in the last part of a sign, its judgment is transferred, and its light is weakened, and its brightness is taken away from it.

30. When a Planet is retrograde in its own domicile and in the 7th degree from the Sun, it is like a man in his own house, and there is someone who has custody of him, namely a man more powerful than himself, and he tries to flee from his fear and cenusre and tyranny.

31. A Planet in a domicile of his enemy is like a man in someone else's house, wherein hatred and anger are already present.

32. A Planet direct in its own domicile, free from weakness and the malefics, signifies the perfection of Questions and of thorough examination, and the completion and effecting [of the thing] for the Querent.

33. When the Moon is separated from Planets, it signifies the past.

34. When the Moon joins together with Planets, it signifies the future.

35. When the Moon is separated from Saturn, lawsuits and sadnesses happen and occurrences of evil and accidents.

36. If the Moon is separated from Jupiter, joy and delight happen, as do friendship, the bestowing of dowries, an addition to the family, and children, and servants, and inheritances, also an addition to wealth, a multiplying of fortunate things, and the dissolving of misfortune and accidents.[243]

37. If the Moon is separated from Mars, quarrels, contentions, and rumors of evil happen, as do pains, bandages, deaths, pouring out of blood, false testimony,[244] fornication of a son, and [a daughter's] harlotry, the removal of jealousy, as in readings, a concession for prohibited masculine intercourse, being put in charge of a woman, drinking of wine made from raisins, and of wine from an intoxicating [source], wounds, amputation of the genitals, making of syrups,[245] those who bring children down from the womb to the vulva,[246] and barrenness of the land.

---

[243]Coley omits most of the words in the significations, as he often does in the following Aphorisms. I have noted a few of his omissions, but not all of them.

[244]Coley ends his translation here, adding "etc." He evidently couldn't construe the remainder satisfactorily, and I can't either. It apparently has errors, and it also appears to be punctuated incorrectly. My translation from this point on to the end of the Aphorism may not be entirely correct.

[245]The Latin has *dispositio syruporum*.

[246]Probably a reference to a surgical procedure used to facilitate a difficult birth.

38. If the Moon is separated from the Sun, sicknesses happen, and anxieties, fear,[247] death, imprisonment, an evil occurrence, and disgrace, pains, accidents, illness, binding, captivity, the cavalry, the army, the people, terrible things.

39. If the Moon is separated from Venus, fornications happen, and jokes, laughter, dancing, singing, silken clothes, betrothals, diversity.

40. If the Moon is separated from Mercury, things happen according to the quantity of the application of Mercury to the fortunes or the malefics.

41. If the Moon is applied to any one of the Planets, it signifies the future according to [the nature of] the Moon's application to the planets.

42. When the Moon is fortunate and laudable in the morning, he who asks on that day about a business matter or about a nativity, [will find that] it will be favorable for him [and] fortunate for every Question.

43. When the Moon is unfortunate or corrupt in the morning, whatever is asked about on that day is not favorable for the Question; and whoever is born does not last long [or] if he does last, he is weak in his life.

44. When the Moon is joined to Saturn, that day is evil for every kind of action and Question.[248]

45. When the Moon is joined to Jupiter, that day is good for every kind of action and Question.

46. When the Moon is joined to Mars, that day is not favorable

[247]Coley stops here with "etc.' and omits the rest of the Aphorism.

[248]Aphorisms 44-73 provide a ready means of judging the suitability of a day for any particular purpose. They could therefore be used to make a quick judgment for a client, or they could be used to prepare a sort of general Daily Guide, based not upon the Client's Sun sign, but rather upon his desire to know what a particular day was good for, or what he should avoid doing or beware of on that particular day.

for any kind of action.

47. When the Moon is joined to the Sun, that day must be disparaged for every kind of action, except for those actions in which someone wants things to be concealed; if someone becomes sick, he will die from that cause.

48. When the Moon is joined to Venus, that day is favorable for every kind of action, particularly when the Question is about marriage and the return of a dissolute woman.

49. When the Moon is joined to Mercury, that day is good for every kind of action, especially bestowing dowries, to meet with[249] writers and stewards; it is good to buy, sell, and make computations.

50. When the Moon is opposed to Saturn, that day must be disparaged for every kind of action; it is not suitable for making a proposal, nor is it profitable.

51. When the Moon is opposed to Jupiter, the day is laudable for every kind of action,[250] especially for those requiring truth, bestowing dowries, seeking agreements, copying things, reckoning interest[251] for stewards; going near or far is good.

52. When the Moon is opposed to Mars, that day is [good for] nothing for any kind of action.

53. When the Moon is opposed to the Sun, that day is not good for any kind of action.

54. When the Moon is opposed to Venus, it is a day laudable for every kind of action,[252] especially for sexual intercourse, and

[249]Here and elsewhere in this *Centiloquy*, the verb *obvio*, which normally means 'oppose' has its late Latin meaning of 'meet with'.

[250]Coley says, "*(Vix Credo.)* [I scarcely believe this.], for oppositions rarely produce good effects." And he omits the rest of the Aphorism.

[251]Reading *fenum* 'interest' instead of *senum* 'of old men'.

[252]Again Coley says, " *(Vix Credo)* [I scarcely believe this.]" And he adds, "The contrary often happens, especially to women."

things having to do with women, for seeking friendships and their associations, good for returning a separated woman; it is also suitable for taking a trip.

55. When the Moon is opposed to Mercury, it is a middling day for every kind of action.[253]

56. When the Moon is in square aspect to Saturn, that day is laudable for every kind of action, especially for meeting with kings, princes, and nobles of the kingdom; it is not good to journey on that day,[254] as the end will be unfortunate; a sick person will die on that day.

57. When the Moon is in square aspect to Jupiter, that day is laudable, especially for determining the truth and asking about marriage, and for access to nobles and to those who rule the kingdom, to found buildings; it is favorable to go out on that day for any exploit; anyone who does something on that day will be glad.[255]

58. When the Moon is in square aspect to Mars, it is not laudable for any kind of action, especially for meeting with the king, for seeking business negotiations; on that day it is not good to marry; a sick person will die, or blood will exude from him, or from a wound.

59. When the Moon is in square aspect to the Sun, it is a day that is disparaged for every thing done, especially for meeting with the king and nobles, and for seeking business and causes with them.[256]

60. When the Moon is in square aspect to Venus, the day is laudable for every kind of action, especially for segregating one-

[253]Coley adds, "*Not for contracts I'm sure*. Make an exception for them."

[254]Coley translates differently: "...Let not Noble and Eminent Men then take journeys; for they will prove ill"

[255]Coley omits most of this Aphorism, but he adds, "but a day when the Moon is in trine to Jupiter is better."

[256]Coley renders it thus, "it is a good day to manage the affairs of great persons," which I think is wrong, but he adds, "but the *Trine* aspect must be preferred before it."

self with a woman, for putting on new clothes, for doing the things of nobility, of sport, of banquets, of marriage, of sexual intercourse, of buying animals or slaves, of going into bathhouses, of adorning, of going out to seek peace and quiet.[257]

61. When the Moon is in square aspect to Mercury, it is a good day for proposing to sell, to buy, to associate with kings, [to make] a computation for a larger house, and to read books [and to study] sciences.

62. When the Moon is in sextile aspect to Saturn, it is a good day, a good day to meet with old men and stewards, to look into one's own causes and business negotiations and personal things, to enter into the disposal of ones things, to make agreement on any settled business negotiation, and to hear a man, and certainly concerning friendship, and to converse with great men, to be associated with their own councils and with others consulted.

63. When the Moon is in sextile aspect to Jupiter, it is a good day to make agreements in meeting on that day with *alcaldos*,[258] to seek judgment of rights, or an account, and for every kind of good action, to marry, to start a trip, to converse with nobles and highly-placed men, to strengthen great things, to restrain the kingdom, to be present at banquets and weddings.

64. When the Moon is in sextile aspect to Mars, the day is good and better what is in it, to meet with generals and providers of arms, to divide up the army [into units], to provide the arms of wars; it is good to be boastful, to receive or share horses.

65. When the Moon is in sextile aspect to the Sun, it is a good day, especially to meet with old men, stewards, and city managers, to meet with kings, the king's principal men, and nobles to undertake negotiations; to put together a kingdom, to advance one's own business negotiations.

---

[257]Again, Coley omits most of the Aphorism, but he adds, "*The reader must be wary of these Aphorisms*."
[258]A Spanish word signifying 'mayors' or 'justices of the peace'.

66. When the Moon is in sextile aspect to Venus, it is a good day for every kind of action, especially for associating with women and youths or castrated men and greater ladies, to seek any kind of friendship at banquets, for the preparation of shows, and to put on new clothes.

67. When the Moon is in sextile aspect to Mercury, it is a good day for jokes and agreement, the computation of dowries, meeting with stewards and providers in every kind of action, and commerce.

68. When the Moon is in Trine aspect to Saturn, it is a good day for every kind of action, especially for meeting with old men, and lords of the kingdom; and it is laudable for the king; on this day, cultivate the land and found buildings, and look into ancient things.

69. When the Moon is in trine aspect to Jupiter, it is a good day for every kind of action, especially for seeking the truth, and for undertaking every kind of good action and the love of God; it is good for meeting with kings, *alcaldos*, and nobles.

70. When the Moon is in trine aspect to Mars, it is a good day for quarrels, hunting, and meeting with kings, it is good to buy animals, to arrange for wars, to send out messengers, to make presentations in war,[259] choosing the order of battle and military companies.

71. When the Moon is in trine aspect to the Sun, it is a good day for every kind of action; on that day it is better to meet with old men and the king, to do work on the land of whatever sort; it is good to meet with princes and nobles, to tie up flags, to put together a kingdom; everything that is done is good.

72. When the Moon is in trine aspect to Venus, it is a good day for every kind of action, and a better day for questions of bonds of marriage; and it is apt for putting on new clothes, to get ready to go

[259]Probably "planning the disposition of the troops on the field."

to *tendas* and causes, [to make] engagements in love affairs, to buy animals and female slaves, to meet with older persons, to look into banquets and marriages, to work on pictures and cases of love affairs.

73. When the Moon is in trine aspect to Mercury, it is a good day to make agreements, to meet with stewards and scribes, to make provision for all things done.

74. Saturn in the ASC makes the question unfortunate for the Querent; if it is retrograde in the chart, it both delays and overburdens, desolates, and shatters [the Question].

75. Saturn in the 10th disturbs the Querent [asking about] land; business transactions are made with good fortune, and corruption occurs in the Question.[260]

76. Saturn in the 7th corrupts the Question and makes corruption of the business of this Question happen, and it adds misfortune to the corruption.

77. Saturn in the 4th in the Question makes good fortune to occur in the beginning of the matter, and [yet] it brings in misfortune to the completion.[261]

78. Jupiter sets free what Saturn binds.

79. Similarly, Venus sets free what Mars binds.

80. When the Moon is separated from a benefic, she sets free what Mercury binds.

81. When the superior Planets are opposed to the Sun, they impedite and corrupt the business, and they impedite questions.

---

[260]The Latin has *Saturnus in 10 terrae quaerentem, sollicitat, negocia fiunt cum fortuna...* I have disregarded the comma after *quaerentem*, and I have supposed it means, 'Saturn in the 10th disturbs someone asking about land...', but I think that either the Latin is corrupt or else one or more words have dropped out. Coley has simply, "*Saturn* in the tenth house, he destroys the most hopeful things, let it be in a Nativity or Question.'.

[261]Reading *completionem* 'completion' rather than *corruptionem* 'corruption'.

82. Saturn in square to the Sun signifies the advancement of the business, and yet it terminates in misfortune and corruption.

83. Saturn in square to Jupiter or in its opposition takes away evil and removes oppression, or it removes its own impression.

84. Saturn in square, opposition, or conjunction Mars prohibits joy, ruins the business, and corrupts or impedites the Question.

85. Saturn in square to Venus, or in conjunction or opposition, completes every Question with something illicit and obedient; and if it is because of the diversity of obedience, I say we are saying it will revert, it takes away shame and it will corrupt.

86. Saturn in square to Mercury, or in opposition, or in conjunction, corrupts the question and ruins the business, and it makes an overflow of superfluity.

87. Jupiter in square to Saturn, or in opposition, or in conjunction, prevents its corruption and changes [the indication] from misfortune to good fortune.[262]

88. Jupiter in square to Mars, or in conjunction, or in opposition hinders its generation and corruption and impression.

89. Jupiter in square to Venus, or in conjunction, or opposition, signifies the completion of the Question and a favorable end for it.

90. Jupiter with Mercury in square, or in conjunction, or in opposition, signifies the perfection of the Question and the settlement of business negotiations and their perfection, and in addition good fortune, and good things.

91. When a malefic Planet is in the ASC and there is a benefic in the second house, the question will be changed[263] from misfor-

[262]This Aphorism is mistakenly numbered 88 in the Latin text, and thereafter the numbering of the Aphorisms is off by one unit. I have corrected the error in this translation.

[263]Reading *quaestio transmutabitur* 'the Question will be changed' instead of *quaerentem transmutabit* 'it will change the Querent'. Cf. Aphorism 92.

tune to good fortune.[264]

92. When there is a benefic in the ASC and there is a malefic in the second house, the Question will be changed from good fortune to misfortune.[265]

93. When a strong Planet is in the ASC, and in one of the angles there is <another> Planet, the question will be changed from good fortune to misfortune.[266]

94. When the Ruler of the ASC is in the combust Way, the business and whatever was asked will be corrupted.[267]

95. When there is a malefic in the 10th or the 4th from the ASC, the strings [of the heart] will tremble, and a misfortune will occur, and [the matter of] the Question will be deprived and cut off.

96. When a malefic is in the ASC and there is a benefic in the 2nd house, the question will be completed, and the business will be diminished with joy and a famous profit.[268]

97. When a malefic is in the ASC and there is a malefic in the 2nd house, the [matter of the] Question will go back, and it will proceed to misfortune.

98. When the ASC is diminished in its ascensions, and its ruler is in a descending house, in the 6th or the 12th,[269] the querent will hope for things that he will not get, and when there is a benefic in the 3rd from the ASC, and a malefic in the ASC, the Querent will have little connection with the business.

---

[264]Coley adds, "Understand the same both in Nativities and questions."

[265]Coley adds, "which must be left to the Astrologers Judgment to determine.'

[266]But it seems that it should read, ". . . in one of the angles there is a <malefic> Planet."

[267]Coley adds, "The Combust way is from 15 d. of ♎, to 15 d. of ♏."

[268]Coley renders this as, "[it] portends the business to answer the Querents desire; but shews him to reap small Gain thereby."

[269]Coley begins, "Few degrees Horoscopical, and the Lord of the ascendant in the ninth or sixth descending. ..."

99. When a Planet happens to be afflicted at the time of its ascension, the Querent gets misfortune at the end of his business, after having brought it to a successful conclusion.[270]

100. [271]When a malefic is received in the angles, it will give and connect the Querent's business, but he will receive from it [only] what he gave himself. The theory of reception is when a Planet is in its own domicile or in the sign of its exaltation, it is free from misfortune.[272]

Note. On p. 345 of his book, after having given his translations of the *Centiloquies* of Ptolemy, Hermes, and Bethen, Coley adds this comment:

[*In these* 300 *Aphorismes is contained concisely the whole Mistery of Astrology; and though they have been already formerly Printed in Almanacks; yet that the Artist might be furnished with them altogether at one view in* English, *I thought it very necessary to insert them here, which I hope will not be taken ill by any, but rather friendly accepted, especially by the true sons of* Urania.]

---

[270]Coley translates the Aphorism thus, "A Planet signifying any matter, evilly disposed at the time of the Querent's first moving the same, Denotes a Vexatious, and Unhappy End to the thing required: if well disposed, judge the contrary."

[271]This last Aphorism is unnumbered in the Latin text due to the numbering error mentioned in the note to Aphorism 87.

[272]This is not the theory of reception! Coley expands on the ideas expressed in this Aphorism thus, "When the Significator of a thing is in Reception, or good aspect of an Infortune in an Angle, the Querents business may then be accomplished; but he usually takes away what he gives hopes of at the last. The knowledge of Reception is, when a Planet shall be in the House or Exaltation of another, and that other in his; and both there free from the beams of the infortunes." But Coley has actually defined *mutual reception* not simple *reception*.

THE PROPOSITIONS OF THE ASTROLOGER ALMANSOR TO THE KING OF THE SARACENS

TRANSLATOR'S PREFACE

This is a translation of the Latin version of *Almansor's Propositions* that was translated from the Arabic in 1136 by Plato of Tivoli with the help of Abraham bar Hiyya. I believe that it was first published, as a supplement to the *Quadripartite* of Ptolemy, by Bonatus Locatellus at Venice in 1493.

THE PROPOSITIONS OF ALMANSOR

My King, I have put together a small compendium of Aphorisms, so that I might satisfy your wishes—a task that I by no means rashly [would have] declined. I ask that you receive [what] I have written with a friendly mind.

1. The disposition of the signs, as I say, is from the beginning of Aries; one of them, namely, is diurnal, another one is nocturnal; one of them is masculine, another one is feminine; one of them is light, another one is heavy.

2. The exaltation of each of the seven planets is said to be in that place in which it substantially suffers from another contrariety. Just as, of the Sun in Aries, which is the fall of Saturn. For the Sun has brightness, Saturn darkness; and as Jupiter in Cancer, in which Mars has its fall, one of which seeks justice, but the other signifies injustice; and so of Mercury in Virgo, which is the fall of Venus, and the one signifies knowledge and philosophy for the Native, but the other one causes joys and whatever is refinement for the body.[273]

3. The signification of the parts of the circle, O Great King, is above one [element] only, just as a sign of the fiery part signifies over fire, but of earthy over earth, of airy over air, and of aquatics over waters.

4. Aries, Leo, and Sagittarius is a hot and dry triplicity, hooked, fiery, diurnal, masculine; and it signifies red choler. Taurus, Virgo, and Capricorn is an earthy triplicity, cold, dry, nocturnal,

[273]An interesting attempt to explain the exaltations. But actually the exaltation signs of the Planets were assigned by the Babylonians for reasons unknown.

and feminine; and it signifies black choler. Gemini, Libra, and Aquarius is an airy triplicity, hot, moist, bloody, sweet, diurnal.[274] Cancer, Scorpio, and Pisces is a watery triplicity, cold. moist, phlegmatic, tasteless, nocturnal.

5. The benefic [Planets] are generous and faithful. The malefic [Planets] are greedy and unfaithful.[275]

6. The benefics will not give anything to anyone with labor. But the malefics, whatever they will give, they will abbreviate and take away.

7. When the two malefics[276] are conjoined, there is a perfected fortune; just as from the conjunction of two benefics; and this is according to what Ptolemy said.

8. Of all the aspects of the planets, the trine and the sextile are better. But the square and opposition are worse.

9. The trine and the sextile of the malefics do not help, just as the square and the opposition of the benefics do not harm.

10. From many significations of life in the *Hyleg*, there is signified much and long life, and also a good intellect and the greatest vigor.

11. Whoever has Mercury in the radix of his Nativity in a domicile of Mars will have bad suspicion and be hurried in his business transactions.

12. Whoever has Mercury in the radix of his Nativity in a domicile of Saturn will be of a great intellect and of lengthy thought, a wise person, and a philosopher.

---

[274]The text mistakenly starts a new paragraph after 'diurnal' and numbers it 4, although the preceding paragraph (beginning with Aries, Leo, etc.) was already numbered 4. But plainly the second paragraph numbered 4 is part of the previous paragraph, so I have combined them

[275]I have replaced the old terms 'fortunes' and 'infortunes' with 'benefics' and 'malefics'. And in this one Aphorism I have inserted the word 'Planets' after each of them, but hereafter I have just used the terms 'benefics' and 'malefics' by themselves.

[276]Saturn and Mars.

13. He will never be a pauper and poor, the ruler of whose Nativity is Jupiter.

14. Whoever has malefics in his ASC will suffer a foul mark on his face.

15. His eyes cannot wander away from a severe impediment, in whose Nativity both of the lights are impedited.

16. He will not lose his sense, in whose Nativity the Moon is helpful to Mercury.

17. When the benefics rule the Nativity, the Native will be generous and lovable. But if the malefics rule, he will be greedy and shunned.

18. The one who has Saturn ruling in his Nativity will be dirty, and the one who has Mars ruling will be stinking.

19. In whoever's Nativity there are malefics in the 8th house, he will die a bad death. But if the benefics are there, he will die easily in his own bed.

20. He will not gather together money and hoard it, unless the ruler of his ASC and the ruler of the 4th are the same planet; just as he will not acquire money or wealth, and he will not live splendidly unless the ruler of his ASC and the ruler of the 10th house are the same.

21. There cannot be affection between two people unless in the radices of their nativities the lights are interchanged, that is, unless one of them is in the place of the other in the Nativity of his comrade. But there will be mutual dislike for those who are born with the opposite [placement] of these, and in signs that are opposite or in square, or with the rulers of those signs being in that mode, or with the lights aspecting each other in that mode.

22. The ruler of the fifth circle,[277] namely Venus, signifies wet

[277] Almansor counts the circles of the Planets from the outermost one down, thus Saturn's circle is 1, Jupiter's is 2, Mars is 3, the Sun's is 4, Venus's is 5, Mercury's is 6, and the Moon's is 7.

spells and rains, just as the ruler of the sixth signifies journeys and changes and travels.

23. He who approaches the king, with the Moon in Aquarius, will not be received, nor will the king pay any attention to him. Moreover, if he approaches him, with the Moon in Pisces, the king will turn his face away from him.

24. There will not be a good diminution of spirit, with the Moon in Gemini, just as it will not be convenient to use fickle procedures, with the Moon in Taurus.

25. The better signs for taking laxatives and purgatives are the water signs, but the better one of them is Scorpio, and the worse one of them is thought to be Cancer.

26. The better sign for association is Leo, but the worse one is Aries.

27. The fixed stars give grand gifts, and they raise people up from poverty to loftiness, which the planets do not do.

28. He will be a perfect physician for whom Mars and Venus are in the sixth. But he will be a good singer whose Mercury is retrograde with Venus in the same sign.

29. The ruler of the fifth circle, namely Venus, when it is in the 1st degree of Cancer, will make rains.

30. When the heavy planets are occidental to the Sun, they will give honesty around the end of life, and vice versa.

31. The one who buys something, with the Moon [posited] from the head of Cancer to the end of Sagittarius, buys dearly, and he sells cheaply; but vice versa when she is [posited] from the beginning of Capricorn to the end of Gemini.

32. It is not a good circumcision [that is performed] with the Moon in Scorpio, just as it is not a good vomiting [that is provoked with the Moon] in Leo.

33. From the nature of that sign, in which Venus is, it will attribute [something] to the Native; for when she is in Leo, she will give love, and in Scorpio sexual intercourse. The reason is because Leo has the heart, Taurus the throat and neck, Scorpio the private parts.

34. Whoever has Mercury in the 12th will be wise and a great philosopher.

35. When the Moon is impedited in the radix of the Nativity, and all the other rulers of the triplicities and the *Hyleg*, but especially the first ruler of the triplicity of the ASC, and there is a malefic in any of the angles, it indicates a short life for the Native.

36. The ruler of the fifth circle, namely Venus, destroys what the ruler of the third circle bestows, and that is Mars. Moreover, the ruler of the second [circle], namely Jupiter, destroys those things that the ruler of the first [circle] bestows, namely Saturn; and this signification is about great things.

37. When all the rulers of the triplicities have fallen, and in any one of the angles there is one of the fixed stars of the first or second magnitude, which is of the nature of the rulers of the triplicities, his rearing will be perfected and he will pass through it.

38. What falls upon the Moon from the malice of Mars, and with her increased in light, is equal to that which falls upon her from the malice of Saturn, with her diminished in light. Moreover, this configuration signifies great things, and he will not be known to be one of the heretics, unless it is from the ruler of the second circle. And the ruler of the fourth circle [namely the Sun] signifies about kings.

39. In elections [made] for kings, the signs [ruled by] the higher planets are recommended, just as in elections [made] for powerless people, the signs [ruled by] the lower planets are recommended.

40. Those things that one of the benefics indicates, [which] are Jupiter, the Part of Fortune, the Part of Giving, and also the 2nd

house, and the one that is its ruler, also the ruler of the triplicity of the Light [of the Time],[278] which has authority, as well as the 10th [house] and the one that is in it.

41. When a significator will be between the ASC and the 10th [house], the [indication] will be [for] days and hours; when [it is] between the 10th [house] and the 7th [house], [the indication will be for] weeks and months; but when it is between the 7th [house] and the 4th [house], [the indication will be for] years.

42. The ruler of the first circle, namely Saturn, when it is in fixed signs, will make mortality and scarcity; moreover, wise men have experienced this.

43. When you want to make an election for someone, and you cannot delay, if the Moon is impedited, put the impediting [planet] as ruler of the ASC.

44. An *ittisal*[279] is better in [both] longitude and latitude; the benefics strengthen the nature of good and reduce the nature of evil; but the malefics act in a contrary manner.

45. The nature of any planet is superior; its action will not cease, unless its contrary [planet] is there.

46. When you are not able to delay an election, you should put the ASC and its ruler in safe [positions], also you should put the names of the fortunes in an angle, and what is better, in the 10th [house]. You should not take note of Mars in travels on water, just as you should not take note of Saturn in travels on land.

47. In travels, the fixed signs are disparaged, but the mobile signs are recommended.

48. When an malefic aspects a significator, and it is retrograde and cadent, and also in a place in which it is *peregrine*, i.e. in none

---

[278]The Sun is the'Light of the Time' for diurnal charts, and the Moon is the 'Light of the Time' for nocturnal charts.

[279]This is the Arabic word *al-ittiṣâl* (spelled *ictisal* in the Latin) 'conjunction', used in the sense of an application that goes to completion.

of its dignities, and in a sign contrary to its own nature, it causes evil for the significator that nothing will be able to cure, but God alone.

49. When Mars approaches the Earth, it will be peaceful with Jupiter; Saturn also will be peaceful thus with Venus; and when the Moon approaches the Earth, and there is an ascension of two parts in one part.[280]

50. The Moon agrees with the Sun when it is increased in light and in number[281] also, every bright star agrees with it.

51. When Mars is debilitated, Venus is strengthened; just as Jupiter, with Saturn weak, lessens the malice of the malefics when it is in any of its own dignities, and vice versa.

52. When a malefic receives a benefic, it will not impedite it, especially moreover when it does not aspect it by square or opposition, and is not in the same sign; the greatest impediment of the planets is when they are in peregrine places.

53. Every planet when it is conjoined in the same point as the Sun will go swiftly; when it comes to it, it will go slowly.

54. Stronger testimony is an *ittisal*[282] of the Moon to a Planet that is in the MC [or] in the ASC.

55. Cancer is the significator over waters that are very moveable, but Scorpio is the significator of waters running in rivers; Pisces also is the significator of immobile waters, such as pools and cisterns.

56. Everything that is made quickly and is destroyed quickly, and which is done repeatedly is signified by Mars.

57. Everything that is done quickly and suddenly destroyed,

---

[280]The Latin has *& est ascensio duarum partium in parte una* '& it is the ascension of two parts in one part'. I do not know what he means by this.
[281]'Increasing in number' means 'swift'.
[282]An *ittisâl* is an application going to completion. See Note 4 above.

and that remains destroyed for a long time, is of the nature of Saturn and Mars.

58. Whoever is of the royal stock, if in the radix of his Nativity there were two Suns,[283] he will be contrary to kings, and he will be separated from them in his actions.

59. Whoever has in the hour of his Nativity the Moon in Taurus in the minute of the ASC or[284] the Sun in Leo in the ASC in the minute of the ASC, will come to a great elevation.

60. Those things that happen in this age are known and investigated from the great strength of the superior significator and from its own elevation, and if it is not, it is sought from the Planet to which all the other Planets. . . .[285]

61. If any one of the Planets signifies anything in the radix of the Nativity, when the division and its ruler come to it, its signification will appear, whether it is good things or evil things.

62. When the ruler of the 10th [house] will retrograde from the *ittisal* of the ruler of the ASC, the Native will not be received by his own lord, when the ruler of the ASC and the Moon are in ascension, and the two benefics similarly are in mutual aspect, the Native will be very strong and powerful, and nothing will escape his command.

63. The accidents of the body are known from the degree of the ASC; from the degree of the Part of Fortune the essence of his personal assets is known. But from the degree of the Moon the essence of his body and mind is apprehended. Moreover, from the degree of the Sun, his health, but from the degree of the MC his personal status and actions are distinguished; give one year to each degree.

---

[283]The Latin has *duo Soles* 'two suns'. This must be a mistake, but I cannot think how to emend it.

[284]The Latin text has & 'and', but the two positions mentioned are mutually exclusive, so I have translated it as 'or'.

[285]Something is missing here (<are applying>?).

64. He will be made fortunate, and he will be of good essence, as well as powerful, whose Revolution of the Year[286] is like his radix, and whose circle[287] is in the same likeness in which it will be in the year of the radix.

65. The daily status of the Native is received from direction, either from the motion of the ruler, or the day to day situation of the Native is received from the motion of the ruler of the sign of the profection of the year set forth in the terms of the ruler of the sign of the *alynthia*[288] of the year to the terms set forth.

66. When Saturn ascends into the higher part of its own circle, that is into its apogee,[289] and when the Moon will be conjoined to the Sun at the end of the month, and she will be favorable to him, it will signify an increase in the thing, which is from the nature of the sign in which it is.

67. When a planet is in the very minute of the degree of the ASC, and the Moon will be conjoined to it in the same place, an act of the nature of that Planet will appear, whether it is good or evil.

68. When any particular Planet is the ruler of any year of the Years of the World,[290] in the very degree of its own exaltation its signification will appear, and a king will be raised up in that region and in the clime in which it will be signified by it.

69. An impediment of Mercury by Saturn must impede the *native's* tongue; moreover, it is worse if it is conjoined corporally.

70. How great will be the fear about a sick person when the lights are under the Earth at the hour of the question.

---

[286]That is, his Solar Return.

[287]By circle' he means the 'circle of houses' or simply the chart.

[288]This is the '*al-intihâ*ˀ of the year'—an Arabic term for a profection from a previous conjunction of Jupiter and Saturn. There were four of these with different periods. Probably the fastest one is meant here. It moves one sign every year.

[289]In the year 2000, Saturn's apogee was at 273°, and since it moves forward nearly 2° per century, from about 320 A.D. to 1998 A.D., it was in Sagittarius.

[290]That is, an Aries Ingress.

71. When both of the Lights are in the same degree of their own exaltation in their own rulership, free from the malefics, the Native will be king of the whole world, and his seed will inherit his lands, and it will be obtained for much time.

72. When the ASC of any Nativity is the MC of the World, and its is namely the sign Aries or Cancer, and that degree which is over the line of the MC is the same degree in which the Sun's or Jupiter's exaltation is, his name will be spread abroad through the whole of the Earth, and his fame will be widespread.

73. These are those that must be adapted, or can be adapted from them, namely the ASC, the Part of Fortune, the Lights, the sign of the new Moon, and their rulers, the sign of the ruler, the hours, as well as the place of the thing sought and its ruler.

74. From the ruler of the exaltation and the divisor, as well as the ruler of the other rays of the divisors, and from the ruler of the year, and also from the change of the Planets in their places, and their aspects, are known the accidents in the Revolution of the Years of the World.

75. When the ruler of the 10th house is in the 8th, it will be feared that the *native's* mother will die from that birth.

76. When the ruler of the 4th house is impedited by the ruler of the ASC, there will be fear about the *native's* father.

77. Nothing is brought forth for the one who is battling when the ruler of the ASC is a malefic[291] and retrograde, or under the Sun beams, which similarly [occurs] if it was in the 7th or with its ruler, the one who first began to fight will win in the battle.

78. A malefic heats up its own place and impedites it; but the Sun heats it up but does not impedite it.

79. Take care lest the king go out to battle when the ruler of the

[291]The Latin text has *in fortuna* 'in a fortune', but I think this is a mistake for *infortuna* 'a malefic'.

ASC is moving toward the ruler of the 7th.

80. When the Sun is with Mars in northern signs, there will be the greatest heat; similarly too, when the Sun will go towards Saturn in southern signs, there will be the greatest cold; and vice versa.

81. When the malefics come to a place appropriate [to them], they will not do any harm, if such a signification was not in the radix of the Nativity; and similarly, the benefics will not produce any good when they do not have such a signification in the radix.

82. The impediments that are going to come will occur in the years of *alynthia,*[292] when the [profection of] the Year of the World comes to the bodies of the malefics. Give a year to each sign.

83. The misfortune due to Saturn is greater when it is in feminine signs; but that of Mars is greater when it is masculine signs.[293]

84. No one should attack a city, whose ruler[294] is the ruler of the ASC of [the Revolution] of the Years of the World.

85. In every election, the circle[295] should be adapted to the nature of the action that is going to be undertaken.

86. An *ittisal*[296] of the Moon from the signs of Venus is not recommended, nor one to Jupiter from the signs of Saturn and Mercury, nor one to the Sun from the signs of Saturn.

87. It is more fortunate when the diurnal Planets are oriental to the Sun, in masculine signs; but the nocturnal Planets [when they are in feminine signs], occidental to the Moon.

---

[292]See Note 288 above.

[293]This is because each Planet is then in a sign whose sex is contrary to the Planet's own nature.

[294]I suppose that he means the *astrological ruler* of the city. For example, London is thought to have its ASC in Gemini, so its ruler would be Mercury.

[295]He means the circle of houses, i.e. the electional chart.

[296]An application going to completion. See the Note to Proposition 54 above.

88. When the rulers of the triplicity of the Light of the Time[297] are diurnal and oriental; in the same way, when[298] the rulers of its triplicity [are] nocturnal [and] occidental, and they will be increased in light, it will be a greater sign of good fortune and happiness.

89. An oriental point signifies young children and the beginnings of each one; but the MC signifies kings and laws, rulers, judges and also the rulers of battles. The 7th also [signifies] old men and dead persons, women too, and enemies, and every pleasantness. Moreover, the Angle of the Earth[299] [signifies] fathers and lands, also the place in which the infant is born. It also shows the hour of death and burial.

90. The Planets that give great riches are three, namely Jupiter, the Sun, and Mercury.

91. Every inception must be feared from that which is of many riches,in all those [undertaken for] moderate things, in which the new Moon or the full Moon was impedited; therefore, beware of this.[300]

92. Every question should be compared to the Nativity; therefore, revolve its years and judge from that.

93. Saturn rejoices in Aquarius, just as Jupiter does in Pisces; Mars also rejoices in Scorpio, just as Venus does in Libra and Mercury in Virgo.

94. Just as *Geuzahar*[301] adds onto the nature of every Planet that is with it, so Cauda[302] iminishes the nature of each Planet that is with it.

95. The stronger signification of the father is from the first

[297]The Latin text has 'day', but I think 'Time' is what is meant.

[298]Reading *eodem* 'in the same way' instead of *ab eo quoque* 'from it also'.

[299]The IMC.

[300]That is, don't expect anything to turn out well when the preceding syzygy was impedited.

[301]A corruption of the Arabic word *al-jawzahar* of Persian origin for the 'Head of the Dragon' or the 'North Node'.

born, and those who follow will give similarly some signification.

96. *Algebuthar*[303] is stronger in the case of life, but the *Alcochoden*[304] is stronger in other things, but the ruler of the rays is stronger than the Algebuthas and they will perhaps be equally strong.

97. When the Native dies before he has passed one day and one night, it will be just like an aborted child, and it will have no signification in anything.

98. Kings whose personal affairs are quickly done, and who quickly avenge themselves, who also quickly accomplish what they desire, are those whose ASC in the beginning of their reign was in a fire sign, and similarly the MC or another angle is in a fire sign or else in an air sign.

99. When the ruler of the ASC is in a good place in the terms of the benefics, and the ruler of the terms is well disposed, being so disposed it will give honesty and a kingdom to him and everything good.

100. When the significator is in the MC, it is directed by the ascension of the direct circle. But in the ASC, by the ascensions of the region; moreover in that which is between one place and another; al-Kindî related this chapter.

101. The cause of good fortune and prosperity is that the ruler of the domicile containing the Sun, the domicile containing the Moon, and the ruler of the domicile containing the ASC are oriental, in angles, and aspecting themselves from good places, that is by trine or sextile.

---

[302]The Dragon's Tail or the South Node.

[303]According to Kunitzsch, this is a corruption of the Arabic *al-jârbakhtârîya* from the Persian phrase *jar bakhtâr*, which is the equivalent of the Greek *chronokratôr*, 'time-distributor'.

[304]This is from the Arabic word *al-kadkhudâh*, which is derived from a Persian word meaning 'house lord', which was a translation of the classical Greek word *oikodespotês* 'ruler' (lit. house-lord).

102. Very powerful is he, for whom the Sun is in the MC in a fire sign, and the Moon is in Taurus, whose aspect will be sinister.[305]

103. They are very rich, and those who abound in riches, and have a great name, for whom the ruler of the 2nd house is in its own exaltation or domicile, moving toward the ruler of the ASC, especially if it is Jupiter.

104. The life of all living things is according to the degree of the Sun and the Moon, and this is given from the highest according to *Astapha*.[306]

105. When there is a benefic in one of the places of Mars, and a significator moves towards it, or it moves towards the significator, it will cut off [the signification], just as a malefic would do.

106. Avert your eyes from a figure in which Mars is in an angle, especially when it is rising in Scorpio.

107. When anyone wants to remove something, with the significator aspecting Saturn, he will remove it violently.

108. If anyone has asked for something, with Caput[307] being in the MC, with Jupiter, and the Moon going to it, or if she is separated from it and going to the ruler of the ASC, or if the ruler of the ASC is going to it, it will not fail that he for a short time gets what he wanted.

109. It is not good for one who is wanting to wage war, nor even for a king to take to the road, with the ruler of the ASC both in the 8th house even if it is in its own exaltation.

110. It is not good for the Sun to be in the ASC at the hour of a war, nor for it to be with the ruler of the ASC unless Aries or Leo is ascending.

---

[305]Here the word 'sinister' refers to the direction of the aspect (i.e. to the left).
[306]I cannot identify this word. Perhaps a reference to an astrolabe.
[307]The Head of the Dragon or the North Node.

111. In that part in which the Sun and the Moon are made fortunate, there will be victory for the Querents, just as if they are made fortunate [when posited] from the ASC up to the MC, or from the ASC down to the 4th, the Querent will win; [but] if [they are] in other places, his adversary will win.

112. The greatest impediment of things ought to be those which are in the power of the sign in which there is a cadent Planet, or one that is retrograde, or in an evil relationship to the Sun.

113. The signs signify bodies, but the planets signify those things that move bodies; the places of the Planets in the circle and their substance, and their location with respect to the Sun, signify work and destruction.

114. In any Nativity in which Jupiter is the receiver of the strengths of all the Planets, and it entrusts them and itself to Saturn, and it is received by him, and both of them are oriental and ascending in an angle, the Native will be great and powerful in this age, as well as a good person, also his own name will grow in the world.

115. Beware lest Saturn is with the ruler of the ASC, or in other exalted places, because it is worse than Cauda.

116. Rumors are true if in the hour of hearing it the angles are in fixed signs, and the Moon and Mercury are in fixed signs, and if the Moon is also separated from fortunes, as well as if there is a benefic in another angle, and when you find it so, this signification can neither be mistaken nor fail.

117. The first hour of the day down to the end of three hours is [the time] of blood; but the second three hours [is the time] of red choler; the third of black choler; but the last three [is the time] of phlegm; similarly too at night. The quarters of the lunar month are in the same mode. And also the quarters of the solar year.

118. The one [in whose chart] the ruler of the ASC is to the right of the Sun, and is always elevated above it, i.e. it has *Dustoria,*[308]

[308]*Dustoria* is from the Arabic *dustûrîya*, which is a transliteration of a Middle

and it has already accomplished its orientality, as they say, will be a friend of kings and powerful men, [and] also rich and [a person] of a great name.

119. Do not be in a rush to judge when the planets are mutually conjoined, unless you first consider the nature of the sign in which they are conjoined—whether it is similar to them or not—for if it is similar to them, you will strengthen the signification.

120. When the ruler of the 11th [house] from the ASC, or the 11th from the Moon, or the 11th from the Part of Fortune is surrounded by benefics, it will be a more fortunate [configuration] than any other.

121. When there are two malefics in the degree of the 4th house, and in the same terms, the Native will be unfortunate and very miserable.

122. When Planets are in angles, those things that are signified by them will appear, whether they are good or evil. The signification of the ruler of the fourth circle is the year, but also the ruler of the seventh circle is the month.[309]

123. When the angles of the ASC are mobile signs, and two malefics are in the angles, the Native will be miserable all of his life.

124. When the Moon is in angles, it will strengthen all of her signification, but especially if she has some authority in the ASC.

125. A rapid change cannot occur unless [it comes] from Mars; just as a great delay [cannot occur unless it comes] from Saturn.

126. The status of all good things is changed from good to evil,

---

Persian word signifying a position of power. See Wright's translation of al-Bîrûnî's *The Book of Instruction in the Elements of the Art of Astrology* (London: Luzac, 1934), p. 299 n.1, who explains the word as being derived from the Persian *dastûr vazîr bûd* 'position of authority'.

[309]The "ruler of the fourth circle" is the Sun, and the "ruler of the seventh circle" is the Moon.

or from evil to good, when the signs and figures of the planets—[namely] those that signify them—are changed from ascension to descension and vice versa.

127. Jupiter and Saturn change things and transform them; and the beginning of the variation will be when they are changed from one triplicity to another in their conjunctions,[310] and from one figure into another.

128. There are two conjunctions that are better, one of which is of the Lights, but the other of the two heavy Planets. Masculine Planets are those that act, but the feminine Planets are those on which they act; and similarly too in the case of the signs.

129. Look at the ascension of a Planet and the degree of its exaltation, and also the degree of its fall, for these are the things that signify men and their works.

130. If you find the significator in angles, masculine, and in a masculine sign, but on the contrary a feminine significator in a feminine sign, it will act on it, that is, the one that is masculine and in a masculine sign will prevail over the other one.

131. Significators are of two modes, namely by their substance and by accident, and the seven Planets will signify in both modes.

132. When two malefics are in conjunction, and the Moon is conjoined to Saturn by latitude, there will be famine and mortality. But if it is conjoined to Mars, kings will be changed, and there will be much spilling of blood, and battles in that place that is signified by the sign, and this does not fail [to happen].

133. From the conjunction of Saturn and Jupiter in mobile signs, a change of status and of the World will be known, and a repetition from their conjunctions in the fixed signs.

[310]Jupiter-Saturn conjunctions occur about every 20 years, and they tend to recur in the same triplicity several times before beginning to occur in the next triplicity, a change that is called a Grand Mutation. A recent example is the conjunction of 29 May 2000 in 23 ♉, which will be followed by the conjunction of 22 December 2020 in 1 ♒—a change or mutation from the Earth triplicity to the Air triplicity.

134. When the two benefics are in conjunction, and the Moon is conjoined to Jupiter in latitude, there will be justice and peace in the land. But if it is conjoined to Venus, there will be liveliness, joy, and bodily health, and prosperity.

135. When Venus and Mars are in the same point with the Sun in the domicile and terms of Venus, the *native's* words will be received by men, and they will not be repudiated by anyone.

136. A hermit and a quasi prophet, as well as those whose words are received, is one in whose Nativity Jupiter and Venus are found in the same degree with the Sun.

137. Famous kings followed in this, and whose commands are not scorned, are those whose Jupiter and the Moon are in the same point ascending to its own apogee.

138. He who possesses the greatest vigor and can do great things is one for whom the Sun in the MC with Saturn will be oriental in a masculine sign.

139. When the Moon and the rest of the significators are remote from the angles, their things will not be accomplished, unless it is done in some way by a journey, and not from anything else.

140. When there is no *ittisal*[311] aspect between any one of the Planets and the significator, and both are in the same circle, and from parallel circles in the nadir or the *anahat*,[312] or in the equinoctial path, it will be better than the *ittisal* aspect.

141. Each Planet has two signs, except the Lights, each of which has only one domicile; since of these, the light of Saturn is darkness, and therefore their domiciles are placed contrary to its domiciles.

142. The benefics are faithful and favorable when they are in the places of their exaltation and direct, and also when they are in-

[311]See the Note to Proposition 54 above.
[312]Not identified.

creased in light.

143. In great and exalted riches of things or their beginnings it is commended that there are Lights in the terms of the benefics aspecting themselves and that the rulers of the terms are of the nature of the things being undertaken.

144. Things of the circle, which embrace everything, and which are the greatest, are seven in number. Namely, the beginning of the creation of men. The greatest years of the Sun, and they are 1413.[313] Moreover, when any beginning ascends in the terms of the ASC of any radical inception, and it finds two minutes of the two heavy Planets.[314] The knowledge of the hour of the year in which it will be conjoined in the signs. The ASC of the Revolution of the Year of the World, also the ASC's of the new Moon and the full Moon that is always made prior to the hour of the Nativity or the question.

145. Nothing good or evil will happen, except when the nature or likeness of the triplicity of the signs of the Planets, which are the significators of that place, is changed.

146. When two climes are changed, the aspects of the Planets and their rays are changed.

147. But when two latitudes are changed, the circles of their circular motion are changed. Moreover, in this chapter nothing is so perfectly stated.

148. The impediments of the two Nodes are worse than those of the inferior and superior Planets.

149. Figures of the circle in questions that are of great similarity are due to the similarity of the mind of the Querent.

150. When Jupiter is direct in Aries, without an evil aspect from

---

[313]This seems to be an error. The greatest years of the Sun are 1461.

[314]I do not understand this sentence. Perhaps the Latin text is faulty. The word *minuta* 'minutes' can also be read as *ininuta* (meaningless), or perhaps it is a misspelling of *initia* 'beginnings'.

the benefics,[315] it will give strength and a kingdom in which no injustice will be done.

[315]Or should we read 'malefics' instead of 'fortunes'?

BIBLIOGRAPHY

Albumasar (Abû Ma<shar)
De magnis coniunctionibus . . .
. . . Augsburg: Erhard Ratdolt, 1495.

. . . Flores Albumasaris.
. . . Augsburg: Erhard Ratdolt, 1488.

. . . Introductorium in Astronomiam . . .
. . . Augsburg: Erhard Ratdolt, 1489.

. . . Albumasaris de revolutionibus nativitatum.
. . . (Greek version.)
. . . ed. by David Pingree
. . . Leipzig: B.G. Teubner, 1968).

Almansor (al-Manṣûr)
Propositiones Almansoris ad Regem Saracenorum.
a supplement to:
Liber Quadripartiti Ptolomei . . .
Venice: B. Locatellus, 1493.

Bethen (ibn Ezra)
Centiloquium Bethen.
a supplement to:
Liber Quadripartiti Ptolomei . . .
Venice: B. Locatellus, 1493.

al-Bîrûnî
The Book of Instruction in the Elements
of the Art of Astrology.
trans. by R. Ramsay Wright
London: Luzac, 1934.

Bonatti, Guido
Liber astronomicus.
ed. by Johann Engel
Augsburg: Erhard Ratdolt, 1491.

Carmody, Francis J.
Arabic Astronomical and Astrological Sciences in Latin Translation.
Berkeley and Los Angeles: Univ. of California Press, 1956.

Coley, Henry
Clavis Astrologiae Elimata . . .
London: Ben Tooke and Thomas Sawbridge, 1676. 2nd ed.

Dictionary of Scientific Biography.
New York: Charles Scribner's Sons, 1970-1980.

[Green, H. S. ?]
The Book of Notable Nativities.
[in a volume containing *A Thousand and One Notable Nativities . . . More Notable Nativities* and *Famous Nativities* by Maurice Wemyss.]
Chicago: The Aries Press, [1943]. repr. in facs. [1-2], v-viii, 130,39,43 pp. 15 cm.

Hermes Trismegistus
Centiloquium Hermetis.
a supplement to:
Liber Quadripartiti Ptolomei . . .
Venice: B. Locatellus, 1493.

Holden, James Herschel
A History of Horoscopic Astrology.
Tempe, Az.: A.F.A., Inc., 1996. 1st ed. paper. xv,359 pp. tables diagrs. 22 cm.
Tempe, Az.: A.F.A., Inc., 2006. 2nd ed. rev. paper. xviii, 378 pp. tables diagrs. 22 cm.

Hübner, Wolfgang
Die Eigenschaften der Tierkreiszeichen
in der Antike.
[The Characteristics of the Signs of the
Zodiac in Antiquity]
Wiesbaden: Franz Steiner Verlag, 1982.

Ibn Ezra, Abraham
The Beginning of Wisdom.
ed. and trans. by Raphael Levy and Francisco Cantrera
Baltimore: Johns Hopkins Press, 1939.

Kunitzsch, Paul
Mittelalterliche astronomisch-astrologische
Glossare mit arabischen Fachausdrücken.
[Medieval Astronomical-astrological
Glossary with Arabic Technical Terms]
Munich: Verlag der Bayerischen Akademie
der Wissenschaften, 1977.

Lilly, William
England's Prophetical Merlin.
London, 1644.

Christian Astrology
Modestly Treated of in three Books.
London: John Partridge and Humphrey Blunden, 1647.
Exeter: Regulus Publishing Co., 1985. 3rd ed. facs. repr. of
1647

Morin, Jean Baptiste
Remarques astrologiques . . . sur le commentaire
du Centiloque de Ptolomée mis en lumière par
Messire Nicolas de Bourdin . . .
Paris: P. Ménard, 1657.
Paris: Retz, 1976. 2nd ed.

al-Nadîm, Abû al-Faraj Muḥammad ibn Isḥâq
The Fihrist of al-Nadîm.
trans. and edited by Bayard Dodge
New York & London: Cambridge Univ. Press, 1970. 2 vols.

Ptolemy, Claudius
Ptolemy's Tetrabiblos or Quadripartite . . .
and the whole of his Centiloquy . . .
trans. by J. M. Ashmand
London, 1822.
London: W. Foulsham, 1917. repr.

Tetrabiblos.
ed. & trans. by F. E. Robbins, Ph.D.
London: William Heinemann;
Cambridge, Mass.: Harvard University Press, 1940.

ΚΑΡΠΟΣ
[Karpós]
[Ptolemy's Centiloquy]
ed. by Emilie Boer
Leipzig: B. G. Teubner, 1961.

ΑΠΟΤΕΛΕΣΜΑΤΙΚΑ.
[Aptotelesmatics]
[the latest and best edition of the Greek
text of Ptolemy's *Tetrabiblos*, with a Latin
Preface and notes]
ed. by Wolfgang Hübner
Stuttgart & Leipzig: B. G. Teubner, 1998.

Ramesey, William
An Introduction to the Iudgement of the Stars
London: Robert White, 1653.

Thorndike, Lynn
"Albumasar in Sadan,"
Isis 45, pt. 1, no. 139 (May 1954):22-32
Cambridge, Mass.: Harvard University.

Vettius Valens
Anthologiae.
ed. by David Pringree
[Greek text with Latin notes]
Leipzig: B. G. Teubner, 1986. xxi,583 pp. 20 cm.

Villennes, Nicolas Bourdin, Marquis de
Le Centilogve de Ptolomee
Paris: Cardin Besongne, 1651.

Printed in the United States
202286BV00002B/106-147/P

9 780866 905787